Fractions

Fractions

A Sliver of the Story

Peter D. Schumer

OXFORD
UNIVERSITY PRESS

OXFORD
UNIVERSITY PRESS

Great Clarendon Street, Oxford, OX2 6DP,
United Kingdom

Oxford University Press is a department of the University of Oxford.
It furthers the University's objective of excellence in research, scholarship,
and education by publishing worldwide. Oxford is a registered trade mark of
Oxford University Press in the UK and in certain other countries

Published in the United States of America by Oxford University Press
198 Madison Avenue, New York, NY 10016, United States of America

British Library Cataloguing in Publication Data
Data available

Library of Congress Control Number: 2024922707

ISBN 9780198916536

DOI: 10.1093/9780198916567.001.0001

Printed and bound by
CPI Group (UK) Ltd, Croydon, CR0 4YY

Contents

Preface

Half a league, half a league
Half a league onward

Alfred, Lord Tennyson (1809–1892), *The Charge of the Light Brigade* (1854)

Mathematics is like a good friend, dependable and nonjudgmental while providing needed comfort and companionship. Mathematics is also ubiquitous and timeless and is ready and available whenever you are. Like great art and music, it can serve as both nourishment and medicine for the soul. The Athenian philosopher Proclus said of mathematics, "She gives life to her own discoveries . . . she awakens the mind and purifies the intellect." Since childhood, I have turned to it in good times and bad. Its study takes patience and perseverance, but the effort only serves to enhance the sense of accomplishment and the rewards of deeper understanding. I hope these pages provide you with some of the enjoyment and even happiness that I derived from creating them.

In one way or another, this book deals specifically with fractions. A fraction is a part of a greater whole. Analogously, I just cover a "sliver" of the story in this modest volume. Half a loaf is better than none. Each chapter serves as an introduction to various aspects of fractions. Like a fine diamond, each face illuminates a different view and perspective. I have interwoven a good deal of biography and history to help humanize the mathematical content. By way of background, I assume little beyond high school mathematics. As the book progresses, the level gradually gets a bit more sophisticated. The last few chapters assume some basic familiarity with both differential and integral calculus. However, I try to provide as much background as possible especially when dealing with infinite series which makes up a lot of the text. I hope that readers will be inspired by a chapter or two to then search for additional sources to continue their journey.

The first three chapters are the most elementary. Chapter 1 deals with the relationships between common (or vulgar) fractions and their representations as decimal expansions. Much of the material should be accessible to a child in primary school. However, there is also some discussion of ancient Babylonian arithmetic based on powers of 60 rather than our decimal base 10. The reader can ponder what is unique about each system versus what is universal within mathematics. Chapter 2 builds on the introductory material by introducing geometric series, a tool which is indispensable for the rest of the book. I include a bit about an intriguing mathematical object called the Sierpiński carpet. Next, Fibonacci numbers and relationships are discussed in Chapter 3, including a proof of Binet's formula and some applications. I hope you share my enthusiasm for them.

The next three chapters provide the basic number-theoretic background required to gain a more thorough understanding of the period length of rational numbers that have repeating, but nonterminating decimal expansions. The mathematical prerequisites are minimal. Chapter 4 covers the essential arithmetic of congruence classes. Chapter 5 serves as a helpful reminder of the details of the Euclidean algorithm useful in finding the greatest common divisor of two numbers. Chapter 6 follows with a good discussion of the Euler phi function and a formula for determining it. The final payoff is a greater understanding of the period length of a given fraction.

Chapter 7 is a brief introduction to matrices, in particular some of the basic arithmetic concerning 2×2 matrices. The material is useful in the following chapter dealing with Farey fractions. Chapter 8 develops many of their useful and beautiful properties, all from scratch. Included is an interesting original result that appears nowhere else.

Chapter 9 serves as a bit of a respite from all the computational and arithmetical aspects considered to this point. It is more concrete and geometric in nature, dealing with slicing pieces of cake, cutting a pizza, or marking up sticks. However, the theme of fractions (part of a whole) still prevails.

The next two chapters introduce separate, but equally interesting and accessible topics. Chapter 10 introduces Egyptian fractions, sums of unitary fractions (all numerators are 1's). It's inspiring to see how such ancient people could create such rich mathematics. In Chapter 11, the notion of rational versus irrational is more formally addressed. The chapter is short and serves as an independent interlude.

The next two chapters deal with continued fractions. Chapter 12 is a rather extensive introduction to them, both finite and infinite. Continued fractions have plenty of applications to both pure and applied mathematics. Here I discuss their use in making good rational approximations to irrationals. There is also a discussion of their use in Huygens' design of a planetarium. Chapter 13 discusses Archimedes' method for determining the circumference of a unit circle. I posit the conjecture that continued fractions played an integral role in his calculations.

The final four chapters deal with series—infinite sums of fractions. Chapter 14 is a "crash course" that serves either as a review or as a pared-down introduction to what is normally covered in a second-semester calculus course. There is a full discussion of the harmonic series and similar series lacking a given digit and the like. Included is also a nice discussion of Euler's historically important work on the sum of reciprocals of squares, a singular result which led to much theoretical development within mathematics itself. Taylor series are introduced in Chapter 15 with an eye on tying up some loose ends from the previous chapter as well as to discuss the arithmetic nature of the constant e. Chapter 16 includes three separate but fascinating problems with fractions. The first one is the Diophantine equation $\frac{x}{y} + \frac{y}{z} + \frac{z}{x} = $ n where one strives to find integer values for x, y, and z for a given value of n. The second is Viete's infinite product that relates the number π with an endless string of radicals. The third, known as the Wallis product, is an equation that relates π to an infinite product that includes all odd numbers in the numerator and even numbers in the denominator.

The last chapter displays some beautiful results due to the mathematicians Brouncker, Euler (again), and the incomparable Ramanujan. For most of the book, I include full proofs or at least partial demonstrations of the material presented. Chapter 17 is meant more for enjoyment's sake alone. The first sixteen chapters are akin to taking a studio art class where you are expected to learn techniques and produce some artwork yourself. Chapter 17 is more like an art appreciation class or a tour of a splendid museum.

After reading the book, I hope you view fractions in a new light and get greater enjoyment from mathematics generally. As George Carlin quipped, "Some people see the glass half full. Others see it half empty. I see a glass that is twice as big as it needs to be."

The book is not intended just as a textbook. However, it certainly can be used as supplemental reading in a more general mathematics, deductive reasoning, or math education course. With this in mind, I've included a gross of problems—144 exercises (with full solutions) for readers to ponder and to solidify their command of the material. Enjoy the book as best you see fit. May its timeless wonders become a part of you and bring you some lasting joy!

Acknowledgments

Although most of the figures drawn were my own, several of them together with most of the portraits used are available within the public domain in such online websites as Wikipedia. However, I must thank the American writer Robert Brault for permission to quote him in the epigraph to Chapter 9 and to the estate of Harold Geneen for the epigraph in Chapter 1 (often attributed to the American sociologist and historian W. E. B. DuBois).

I wish to express my sincere appreciation to the excellent professional staff at Oxford University Press. In particular, a big heartfelt thank you to acquisitions editor Dan Taber, who encouraged, gently prodded, and supported me throughout all phases of the writing process. Also special thanks to senior project editors Giulia Lipparini and Jodie Keefe for their kind help in the latter stages of writing and production. Special thanks to the anonymous reviewers whose thoughtful comments have led to noticeable improvements in the final manuscript. I would also like to express my appreciation to my project manager Karthiga Ramu and copy editor Charles Lauder. Finally, and most importantly, I especially want to express my deepest gratitude to four generations of women who will always be a big part of me—Evelyn Horwitt Levine, Janet Ellen Schumer, Lucy Harding Schumer, and Amy Harding Schumer. I dedicate this book to them.

Chapter 1

Terminating and Repeating Decimals

> When you have mastered the numbers, you will in fact no longer be reading numbers, any more than you read words when reading a book. You will be reading meanings.
>
> **Harold Geneen (1910–97), *Mangaging* (1984)**

Anyone who has sliced a birthday cake knows something about fractions. These pages tell a little bit about different types of fractions—common fractions, decimal fractions, continued fractions, finite and infinite sums of fractions, and other variants. For many of us, whether we loved math in school or not, studying the arithmetic of fractions might have seemed repetitious. What I hope to convince you is that fractions can be interesting and contain delightful and surprising patterns. Even a simple common fraction like 1/81 has an intriguing pattern in its decimal expansion, namely

$$\frac{1}{81} = 0.01234567\ldots.$$

(I usually place the integer 0 in front so that the decimal point isn't mistaken for a sentence period.) There is a plethora of fascinating ideas and insights within the world of fractions which reflects the greater wealth of beautiful patterns and structures that permeates the larger mathematical world.

Fractions are ubiquitous in the real world and being able to deal with and manipulate them is an important topic in elementary education. Backing up a bit, the *natural numbers* (or counting numbers or whole numbers) are the numbers that begin 1, 2, 3, 4 which we use to enumerate any collection of objects. These are the first numbers with which civilizations throughout the world have had to grapple since ancient times. The ability to group them together by adding and multiplying soon followed. Four apples plus three more make seven. Three bunches of six bananas totals eighteen. Fortunately, the natural numbers are *closed* with respect to addition and multiplication—meaning that the sum or product of any two natural numbers is another natural number. But subtraction is a bit trickier. For most practical problems, one only needs to subtract a smaller number from a larger one. For example, if I have twelve dollars and spend four dollars, then I'm left with eight dollars. The number zero needs to be appended to our list of numbers if I want to discuss spending all my money. And even then, subtraction is not closed within our set of numbers. We extend our set to include all the *integers.* . . . , −3, −2, −1, 0, 1, 2, 3, . . . which extend infinitely in both directions. The natural numbers constitute the *positive integers.* Unlike the natural numbers, the full set of integers is closed with respect to addition, multiplication, and subtraction. Occasionally I'll use the word integer to refer to positive integers when the context makes it clear. But if we wish to also include division, we need to extend our conception of number even further to include *fractions.*

Some advanced societies never fully developed a notation for fractions even if they dealt with deep scientific and mathematical challenges that seemed to require their use. For example, in the first millennium CE, the Maya of the Yucatan peninsula, neighboring areas of Mexico, and northern Central America had a very highly developed numbering system, an extensive writing system, a splendid artistic and architectural tradition, and a well-organized government and harmonious civic structure. They made remarkably accurate astronomical observations of the "movements" of the sun and the moon and of the planets Venus and Mars. From these observations, they were able to create highly accurate calendars important both for their religious practices and for practical matters of crop planning and food distribution. At Palenque, their estimate for the average length of a month (lunar cycle) was based on their observation that 81 new moons took 2392 days. But there is no Mayan way to express fractions or to rewrite them as decimals. They dealt (quite capably indeed) with just whole numbers. Today we would say that their approximation for the length of a month is 2392/81 or approximately 29.53086 days. Current measurements give a value closer to 29.53059.

The modern perspective is to define the *rational numbers* as the set of numbers of the form *a/b* where *a* is any integer and *b* is any natural number. The great thing about the rationals are that they are closed with respect to addition, subtraction, multiplication, and even division as long as we don't divide by zero. (We can think of subtraction as being addition of negative numbers and division as being multiplication by a non-integer rational.) I am placing the rationals in some greater context to give you a good mental picture of where various numbers are situated. There are some more technical matters that I'll gloss over for now. One issue is that every rational number can be expressed in a myriad of ways. For example, the fraction 1/3 can also be written as 2/6 or 3/9 and so on. In fact, it can also be expressed as $\sqrt{12}/\sqrt{108}$ where neither numerator nor denominator is an integer. But it's all the same rational number 1/3 (when written in *reduced form* where no common factors can be removed from the numerator and denominator).

I'll use the word *fraction* to be synonymous with rational number. Further, decimal fraction will mean the decimal expansion of such a rational number. By the way, the English word *fraction* comes from the Latin word *fractus* which means "to break." According to Steven Schartzman in *The Words of Mathematics*, the words fragile, fragment, and even diffraction have a similar etymology. The ancient Greeks had a more geometric understanding of number and thought of fractions as proportions or *ratios* between lengths of line segments or areas of rectangles. Our *rationals* are just numbers that can be expressed as ratios. For millennia, it seemed that this was it and that there was no need to extend the concept of number any further. In the writings from ancient Egypt and Babylonia, mathematical techniques were presented via specific problems with cookbook-like descriptions rather than giving fuller explanations with more general methods of solution. And a fairly accurate approximation to the actual answer was not distinguished from an exact solution. Hence there was little need to question whether every length of rope (for example) could be measured completely accurately using some rational number of given units as long as a good rational approximation was at hand. One can always find a fraction that is accurate enough for any practical purpose. There is more to the story, of course, which we'll save for later chapters. The set of *real* numbers constitute all possible points on an endless number line. You might know that not all real numbers are rational. We will have much more to say about this in later chapters.

For now, I'll only assume that you can handle basic arithmetic with fractions. For example, if you have half of a pie and someone gives you another third, then you have $1/2 + 1/3 = 5/6$ of a pie which you can compute by first finding a common denominator of 6. Similarly, if you have a recipe for five people which requires 1/2 pound of sugar, but you just want to make the recipe for two people, then you will need $1/2 \times 2/5 = 2/10 = 1/5$ pound of sugar. Having the skill to handle such situations is very useful but not especially interesting.

Let us begin with the reminder that you can work out the decimal form of a common fraction by simply performing long division (or using a calculator to do the work for you). This should be straightforward, but let's take a short spin. For example, refresh your understanding by checking the decimal forms of the common (or *vulgar*) fractions

$$1/4 = 0.25$$
$$1/125 = 0.008$$
$$1/3 = 0.33\bar{3}\ldots$$
$$7/3 = 2.33\bar{3}\ldots$$
$$1/7 = 0.142857\overline{142857}\ldots$$
$$2/7 = 0.285714\overline{285714}\ldots$$
$$1/81 = 0.012345679\overline{012345679}\ldots\ldots$$

(The bar over a set of digits means that they repeat ad infinitum. A fancy word for this bar is *vinculum*.) Note that 1/4 and 1/125 had terminating decimal fractions while the rest were nonterminating. The fraction 1/3 has a repeating decimal representation that repeats from its onset. The fraction 7/3 is simply 6/3 + 1/3 = 2 + 1/3. We just place a 2 in front of the decimal point and copy the expansion of 1/3. There's not much additional interest with such mixed fractions, so most of our attention will be on fractions strictly between 0 and 1. The fraction 1/7 and 2/7 also have repeating decimal expansions of length 6 and have somewhat interesting and related expansions. Finally, 1/81 has a nonterminating expansion which starts interestingly with consecutive counting numbers but seems to go a bit haywire after a while. Even so, it has a nice repeating pattern as well.

What about moving from terminating decimals back to their common fraction representation? In this case, just divide by the appropriate power of 10. For example, 0.24 is the same as 24/100. This can be further reduced by dividing out any factors common to the numerator (number on top) and denominator (number on the bottom). Thus,

$$0.24 = 24/100 = 12/50 = 6/25,$$

which we'll call its reduced form since 6 = 2 × 3 and 25 = 5 × 5 are relatively prime (i.e., have no common factors larger than 1). Try the decimal 0.325. Here you'll get

$$0.325 = 325/1000 = 65/200 = 13/40.$$

Easily done, right?

Next, can we find the common fraction equivalent to a repeating nonterminating decimal? There are a couple ways to proceed, but the simplest is the following. To get going, multiply the fraction by the appropriate power of 10 corresponding to its repetition length and then subtract the original fraction from that. For example, let's consider the fraction $0.246246\overline{246}...$, which has repeating pattern 246. Let's call this fraction x. So

$$x = 0.246246\overline{246}....$$

Multiply by the next power of 10 bigger than 246, namely 1000 = 10^3. This just moves the decimal place over three spaces:

$$1000x = 246.246\overline{246}....$$

Now subtract x from $1000x$. We get

$$1000x - x = 246.246\overline{246}... - 0.246\overline{246}....$$

$$999x = 246.$$

The decimal part has apparently disappeared. It's now a simple division to obtain $x = 246/999$.

If we wish, we can simplify by removing common factors.

$$x = 246/999 = 82/333.$$

We now ask (and then answer) a couple of natural and interesting questions. Rather than working out endless exercises, let's investigate some real mathematics. First, can we somehow describe or characterize those fractions that have terminating decimal expansions? Obviously, there are an infinite number of them, so we can't just make a short but complete list. To simplify matters a bit, again let's assume that the fraction lies between 0 and 1. Larger fractions are just mixed fractions made up of an integer plus such a fraction. For example, 6.25 is just 6 + .25 or 6 + ¼. We can think of all fractions ending in .25 as being ¼ fractions. How can we go about analyzing all such fractions?

List them by number of digits. The one-digit terminating fractions are 0.1, 0.2, 0.3, . . ., 0.9. Recall, for example, that 0.1 = 0.10 = 0.100, etc. We can always append additional zeroes, but the decimal terminated after the first digit. These are the fractions

$$1/10, 2/10, 3/10, ..., 9/10,$$

namely all the positive fractions less than 1 with a 10 in the denominator. There are nine of them. Of course, many can be further simplified so that our list becomes

1/10, 1/5, 3/10, 2/5, 1/2, 3/5, 7/10, 4/5, and 9/10.

Since some fractions have been reduced, their denominators are no longer the number 10 but rather an integer (larger than 1) that divides 10. If you wish, we could include $10/10 = 1 = 1/1$ to get 10 fractions in all with all denominators still dividing 10 (including the "trivial" denominator of 1).

Next, let's consider the decimal fractions strictly between 0 and 1 that terminate after two digits. We can list all the decimals that terminate *by* the second digit by writing them all out as we did above. We get

0.01, 0.02, 0.03, 0.04, . . . , 0.99.

Note that these include our earlier list since for example, $0.20 = 0.2$. Furthermore, we can reduce many of them to obtain

1/100, 1/50, 3/100, 1/25, 1/20, 3/50, 7/100, 2/25, etc.

There are ninety-nine such fractions in all (nine that terminate after one digit plus an additional ninety that include a nonzero second digit before terminating). The fractions comprise all positive decimals less than 1 that have a denominator dividing the number 100. Of course, $100 = 10^2$.

The pattern is becoming fairly clear. Let's highlight our observation.

Observation 1: The decimals that terminate with at most n digits (n being any positive integer) all have denominators that divide 10^n.

For example, fractions with terminating decimals of at most three digits in length must have denominators dividing $10^3 = 1000$. But 10^n can be factored as $10^n = 2^n 5^n$. This is called its prime decomposition since 2 and 5 are prime numbers which cannot be further factored.

Hence, the positive integers that divide 10^n must have at most n factors of 2 and n factors of 5 and no other prime factors. For example, once reduced, the fractions that terminate with at most three digits will have one of the following denominators: 2, 4, 8, 5, 10, 20, 40, 25, 50, 100, 200, 125, 250, 500, or 1000. These are all the possible divisors of 1000 (larger than 1). Double check that I didn't miss any of them. But again, if we don't care to reduce our fractions, then we can just list all such fractions with denominator 1000.

For a final example, let's go the other way. How long is the decimal fraction that represents 7/160? Naturally, we can divide and then look to see the answer. But to just answer the question, we only need to factor the denominator. Here

$$160 = 32 \times 5 = 2^5 5^1.$$

The larger of the two exponents is our answer, namely five digits. Again, the reason for this is that $2^5 5^1$ divides evenly into $2^5 5^5 = 10^5 = 100{,}000$. And we get a nice bonus. Since

$$2^5 5^5 / 2^5 5^1 = 5^4 = 625,$$

we have that $1/160 = 625/100{,}000$ and that

$$7/160 = 7 \times 625/100{,}000 = 4375/100{,}000 = 0.04375.$$

Next let's look a bit at repeating but nonterminating decimal fractions like $0.33\bar{3}\ldots$ or $0.142857\overline{142857}.\ldots$ You probably know the common fractions that are equivalent to them. All rational fractions have either a terminating decimal expansion (which we've just discussed) or have a nonterminating but repeating pattern in their decimal expansion. Let's get a good sense of why this is the case.

Consider any reduced common fraction a/b—meaning that a and b are relatively prime. If the prime factorization of b consists solely of factors or 2 and/or 5, then we know that a/b has a terminating decimal expansion. Otherwise, it doesn't terminate. Let's pick an example of a fraction for which you probably don't know the decimal expansion off the top of your head, say 2/13. To determine its expansion by hand (using long division), we divide 13 into

2.00000 . . . . Confirm that it begins with 0.153, etc. Here you'll want to have paper and pencil handy.

Each step of the way, we divide 13 into our new number n to find the greatest integer k such that $13k$ is less than or equal to n. We then find $n - 13k$ and then append 0 to the right of the difference (obtaining $10n - 130k$). Next, repeat this process. This is the essence of long division.

In our example, we begin by dividing 13 into 20 (actually 2.0) to get a quotient of 1 and remainder of 7. We now know that 2/13 is 0.1 plus something. $20 - 1 \times 13 = 7$ as we noted above. Now append 0 to 7 to get 70. Now repeat. The number 70 divided by 13 is 5 plus change (i.e., with remainder 5). So $2/13 = 0.15 \ldots$ something. This time $70 - 13 \times 5 = 5$. We multiply it by 10 (or equivalently tack a 0 on the end of 5) to get 50 and then continue. Eventually, we get $0.1538461 \ldots$ something. At this point, our remainder is 2, the same remainder that we had at the beginning of this process. Hence the same digits will now repeat, in fact ad infinitum. So

$$2/13 = 0.153846\overline{153846}.$$

Note that there are only twelve different nonzero remainders possible when we divide by 13 (we won't get a remainder of 0 since that would imply a terminating decimal expansion). So eventually we must get a repeated remainder, and at that point we have the full decimal pattern that will be repeated. Of course, this will work with any rational number since the denominator is a finite number which limits the number of possible remainders. What an awesome result! So to recap, any rational number will have either a terminating decimal expansion or one that will repeat after a while (in fact a short while if the denominator is not too big).

Let's dig a bit deeper. What fractions repeat after just one digit, i.e., have a single repeating digit? Obviously, they are the fractions

$$0.11\overline{1}\ldots, \ 0.22\overline{2}\ldots, \ 0.33\overline{3}\ldots, \ \ldots, \ 0.99\overline{9}\ldots.$$

You can confirm that the first decimal fraction is 1/9 by dividing 1 by 9 or by using the same algebraic technique we used to study the fraction $0.246246\overline{246}\ldots$. (It can also be manipulated as a *geometric series* which we will discuss more fully in Chapter 2.) The next decimal expression is twice as large and hence $0.222\ldots = 2/9$. The next is $3/9$ or $1/3$. In all, there are just nine such fractions. Once reduced, they are the nine fractions less than or equal to 1 having denominators that divide the number 9.

The last decimal seems a bit curious. From the pattern we developed, it must be $9/9$.

But $9/9 = 1$. Given that $1 = 1.00\overline{0}\ldots$, could $0.99\overline{9}\ldots$ be another representation of the number one? The answer is yes, though at first you might erroneously think that it's the "number just before 1." In fact, there is no distinct number just before one (since the real numbers are said to be *dense*). Another quick way to confirm our assertion is to let

$$x = 0.99\overline{9}\ldots.$$

Then

$$10x = 9.99\overline{9}\ldots$$

and upon subtracting x from $10x$, we get that $9x = 9$ (since all the digits to the right of the decimal point subtract out). Now divide by 9 to obtain $x = 1$.

This is a key part of a bigger picture. Any terminating decimal can be replaced with a nonterminating decimal where the last nonzero digit n of the terminating part is replaced with $n - 1$ to which we append an infinite number of 9's. For example,

$$1/4 = 0.25 = 0.249\overline{9}\ldots$$

$$19/500 = 0.038 = 0.037\overline{999}\ldots.$$

So our terminating decimals can also be viewed correctly as nonterminating repeating decimals (after some initial digits).

Moving on, which fractions have repeating decimals of length two? Again, for simplicity's sake, we'll just tackle those that immediately terminate without any other "lead-in" digits. These are the fractions

$$0.01\overline{01}\ldots, \ 0.02\overline{02}\ldots, \ 0.03\overline{03}\ldots, \ldots, \ 0.11\overline{11}\ldots, \ldots, \ 0.98\overline{98}\ldots,$$
$$0.99\overline{99}\ldots.$$

There are ninety-nine such fractions if we include those that repeat after one digit for which we have already accounted. (Hence there are 90 new ones.) You can readily verify that the first number is 1/99. The rest are 2/99, 3/99 = 1/33, 4/99, 5/99, ..., 99/99 = 1. Some can be reduced of course. But in any event, the fractions include all those that have denominators dividing $99 = 3^2 \times 11$ (and no others). So it must include 4/9, 5/11, and 13/99, but not include 1/6, 2/7, 5/17, for example.

Taking it one step further, which fractions have repeating decimals of length three? These are the fractions

$$0.001\overline{001}\ldots, \ 0.002\overline{002}\ldots, \ 0.003\overline{003}\ldots, \ 0.999\overline{999}\ldots,$$

which are of course 1/999, 2/999, 3/999 = 1/333, etc. They comprise all the fractions which upon being reduced have denominators dividing 999. Since $999 = 3^3 \times 37$, we have some interesting consequences. For example,

$$1/27 = 1/3^3 = 0.037\overline{037}\ldots$$
$$1/37 = 0.027\overline{027}\ldots,$$

which make a pretty pair. Next, you might want to use a calculator to look at 1/27,027 and 1/37,037 to behold something similar.

More generally, we can now carefully state our second main result.

Observation 2: The fractions having immediate repeating decimals of length n are those with denominators dividing 99 . . . 999 = $10^n - 1$, the integer consisting of n 9's.

For example, let's consider the fraction 1/101. The denominator 101 does not divide evenly into 9 or 99 or even 999. However, the number

$$9999 = 9 \times 1111 = 3^2 \times 11 \times 101.$$

So 101 does divide 9999, and in fact $9999/101 = 3^2 \times 11 = 99$. Hence

$$1/101 = 99/9999 = 0.00990099\ldots,$$

which repeats after every fourth digit. Note that 1/99 is a related decimal. We have

$$1/99 = 101/9999 = 0.0101\overline{0101}\ldots.$$

Of course, this is the same as $1/99 = 0.01\overline{01}\ldots$, which repeats after every other digit.

Let's consider, 1/271. This looks difficult until we note that

$$99{,}999 = 3^2 \times 41 \times 271.$$

Therefore,

$$1/271 = 3^2 \times 41/99999 = 369/99999 = 0.00369\overline{00369}\ldots.$$

By the way, Observations 1 and 2 can be combined. Suppose we are interested in the decimal expansion of $1/n$ where $n = 2^a 5^b r$ where r divides $10^m - 1$. In this case, $1/n$ will eventually have a repeating decimal of length m, but it won't necessarily start immediately (i.e., it won't have a *purely* repeating decimal). Instead, it will have an initial set of digits of length the larger of the integers a and b followed by its repeating pattern. The integer $12 = 2^2 \times 3$, and $1/12 = 0.08\overline{3}\ldots$ has a repeating pattern that begins after its two initial digits. For another example, consider the fraction 1/35. The integer $35 = 5 \times 7$. In this case,

$$1/35 = 0.0\overline{285714}\overline{285714}\ldots$$

with an initial single decimal digit of 0 followed by the infinite repetitive pattern. You may have noticed that $1/35 = 2/70 = \frac{1}{10} \cdot \frac{2}{7}$. Check out some simple examples on your own to double check how this works. The Exercises are a good place to start.

Until now, we have been working completely in base 10 with its concomitant decimal expansions. This is only natural since most of the modern world deals solely with this system. However, mathematical conceptions developed throughout the world in endless, varied, and wonderful ways. Counting was usually *object specific*. The Aztecs of central Mexico counted via stones. One was "one stone," two was "two stones," and so on. Four birds would be expressed as four stones of birds. The Nicie people of South Pacific use numbers equivalent to "one fruit," "two fruit," etc. Some native people of Java in Indonesia count via "one grain," "two grains," and so on. To math historians, this is known as the "concrete" stage of counting.

The English words one, two, three, four, etc. were probably all object-specific at one time as well, but their meanings have long been lost. It is known that "five" relates to "hand" and that "eleven" and "twelve" may have meant "one over" and "two over" as in bending fingers beyond ten. Vestiges of the "tw" in two still remain in the words twist and twine (curling two strings together), twig (where a tree branch divides) and twilight (the time *between* day and night.) By the way, the *The Real Mother Goose* nursery rhyme collection contains this familiar refrain:

> Hickory, dickory, dock!
> The mouse ran up the clock;
> The clock struck one,
> And down he run,
> Hickory, dickory, dock!

The words hickory, dickory, dock are believed to be old folk-words for the numbers 8, 9, 10. Similarly, eeny, meeny, miny, mo might have been common words for 1, 2, 3, 4.

As civilizations advanced and the need to conceive of ever larger quantities, counting was systematized, and numbers were grouped together into common bases (or radices). No doubt due to the number of fingers on two hands, base 10 was a popular choice. But there were lots of exceptions. Some ancient Syrians as well as ancient

Chinese used a binary (base 2) system. A few Native American tribes used a ternary (base 3) system; in Siberia there was a quinary (base 5) system. The duodecimal (base 12) system was also used—often by counting using the joints and knuckles of four fingers. The Huli language of Papua New Guinea uses base 15. And the vigesimal (base 20) system was used by the Ainu of Hokkaido (northern Japan), native Greenlanders, and nearly 10% of Native American tribes. It was especially well developed by the Maya people of Guatemala and the Yucatan Peninsula used in their calendar and astronomical observations. Vestiges of a vigesimal system still exist in modern French where 80 is "quatre vingt."

Much of what's been covered in this chapter could be modified to other bases as well. Let's take a brief look at one very ancient number system quite distinct from our own.

Between 5000 and 1000 BCE many cultures and societies lived and competed along the plains between the Tigris and Euphrates rivers, then known as the "Fertile Crescent," though now largely desert. At the time, the two rivers did not meet. Towns and cities were formed as animals and fowl were domesticated and crops and cereals were cultivated. The societies included the Phoenicians, Akkadians, Hittites, Persians, Assyrians, and many others. The early Greeks referred to the area as Mesopotamia, meaning "land between the rivers," or as Babylonia. Babylon was its main city located about 50 miles south of present-day Baghdad along the Euphrates River.

One common tie among these many cultures was the use of *cuneiform* (wedge-shaped) writing on durable clay tablets. Originally the writing proceeded from top to bottom, but later from left to right. The tablets were created by using a wedge-shaped reed stylus on wet clay followed by baking the tablets in ovens or simply letting them sit out in the hot sun to dry and harden. They are practically indestructible. Over 50,000 such tablets have been unearthed in the region around Nippur alone—some over 4000 years old. Over 300 of these are of some mathematical significance. Of special interest to us here is that their number system was a sexagesimal one (base 60)!

The Babylonians used a thin vertical wedge for the number 1 (which we approximate as a v) and a somewhat thicker horizontal

wedge to denote 10 (which we write as a <.) With these two symbols they could write out the numbers from 1 to 59. For example,
the number 12 is written as <vv and 46 as <<<<vvvvvv. The
contemporaneous ancient Egyptians needed different symbols to
distinguish 1 from 10 from 100 from 1000, etc. Each power of 10 had
its own special symbol. However, with just two symbols, the Babylonians created a primitive place-value system. Our modern decimal
system is a place-value system that uses 10 symbols. We can readily
distinguish between 23 and 32 as well as 27 from 207. The Babylonian system was not as fully developed. Spacing was used somewhat,
but there was still some ambiguity when context was not clear. So
<vvv might mean 13 or possibly 130 or maybe 103. The idea of a
cipher or zero did not exist at this time. (Note that our modern zero
actually has two meanings—as a symbol representing nothing and
as a place holder between numbers.) Eventually (around 300 BCE) a
limited zero was used instead of a space between nonzero entries of a
number, but not as a terminal digit. So the numbers 72 and 702 could
be distinguished, but not 72 and 720. Context was key. And there was
no sexagesimal point corresponding to our decimal point. There was
no written distinction between say 32 and 3.2. Even so, the Babylonians were quite adept at arithmetic, handled addition, subtraction,
multiplication with great facility and even made good approximations for square roots. Most importantly, the Babylonian place-value
system allowed them to ably handle arithmetic problems involving
fractions as easily as those without. Further, the high divisibility of
the number 60 by many smaller numbers allows for its arithmetic
flexibility. And we all live with the vestiges of this ancient system in
our clocks (hours, minutes, seconds), global measurements (degrees
and minutes of latitude and longitude), and even degrees of a circle.

How would the Babylonians write the fraction ½ in sexagesimal?
In base 10, we rewrite ½ as 5/10 and hence ½ = 0.5. In sexagesimal,
we need to rewrite ½ with a denominator that is a power of 60. In
this case, ½ = 30/60 and so ½ = 0.30 in sexagesimal. Combining
some elements of cuneiform writing, we denote this by ½ = 0;<<<
where we use our modern symbol for zero and a semicolon for a
sexagesimal point.

What about 1/20? Rewrite the fraction in terms of powers of 60. Here $1/20 = 3/60$ and hence $1/20 = 0;\text{vvv}$ in sexagesimal notation. As a greater challenge, try to determine the sexagesimal representation of 7/120. We need to go to the next sexagesimal place. The number

$$\frac{7}{120} = \frac{7}{2 \cdot 60} = \frac{30 \cdot 7}{30 \cdot 2 \cdot 60} = \frac{210}{60^2} = \frac{3 \cdot 60 + 30}{60^2} = \frac{3}{60} + \frac{30}{60^2}.$$

Hence, $7/120 = 0;\text{vvv} <<<$ (with a slight space between the 3 and the 30).

By Observation 1, in base 10, the decimals that terminate with at most n digits all have denominators dividing 10^n. Analogously, in base 60, the sexagesimals that terminate with at most n sexagesimal "digits" all have denominators dividing 60^n. The prime factorization of 60 is $2^2 \cdot 3 \cdot 5$. So whereas denominators of terminating decimal fractions need to consist of factors of 2 and 5 only, denominators of terminating sexagesimal fractions can have factors of 2, 3, or 5.

Just as we did with decimal (base 10) expansions, it's natural to wonder how to convert a repeating but nonterminating sexagesimal expansion as well. The Babylonians didn't actually deal with this since they were interested in fractions that terminated. If a fraction didn't have a terminating sexagesimal expansion, they would use a close approximation to the actual answer. Infinite sums were not understood. In practice, this was probably good enough. But let's see what we can do.

Consider the repeating sexagesimal expansion

$$x = 0; << \text{v} << \text{v} \overline{<< \text{v}} \ldots$$

We wish to determine the common fractions represented by x. As we did before, we multiply x by an appropriate factor to move the pattern over to the left. This will enable us to then remove the portion to the right of the sexagesimal point. Rather than multiplying by 10, here we multiply by 60. We obtain

$$60x = << \text{v}; << \text{v} \overline{<< \text{v}} \ldots.$$

Thus,

$$60x - x = 59x = \text{<< v} = 21.$$

Hence, $x = 21/59$. Nothing to it!

For some problems, the ancient Babylonians needed a rational approximation to an irrational number (which we will discuss in greater detail later in the book). One notable example is given in the Yale Collection problem #7289. Without explanation, the value of

$$\text{v; << vvvv <<<<< v <}$$

is given as an approximation for the number $\sqrt{2}$. Is this a good approximation? How did they find it? Let's investigate.

Rewriting in our Hindu-Arabic numeral system, the value given is

$$1; 24, 51, 10 = 1 + \frac{24}{60} + \frac{51}{3600} + \frac{10}{216,000}.$$

The last fraction simplifies to $\frac{305,470}{216,000} = 1.41421296296\overline{296}\ldots.$

The actual value of $\sqrt{2}$ is $1.414213562\ldots.$ So their value is a very slight underestimate with error less than 0.0000006. Quite extraordinary!

Their method of discovery is not recorded. But here is an explanation that fits the facts well. Notice that if x is a good approximation to $\sqrt{2}$, then $2/x$ is another good approximation since $x \cdot \frac{2}{x} = 2$ and $x \approx$ (is approximately) $2/x$. If x is a bit over $\sqrt{2}$, then $2/x$ will be just a bit under and vice versa. So begin with an approximation x to $\sqrt{2}$, take the average of x and $2/x$ and we'll have an even better approximation. Then use the result as our new x and continue recursively. Each step of the way, we'll get a better and better approximation to $\sqrt{2}$. This is a technique that ancient scholars were fully capable of utilizing and is consistent with ancient Greek mathematical practices centuries later. In this case, rather than x, let's call our first approximation a_1. We choose

$$a_1 = 2 \text{ and then } 2/a_1 = 1.$$

Taking their average (arithmetic mean) we get

$$a_2 = 3/2 \text{ and } 2/a_2 = 4/3.$$

Their average is

$$a_3 = \frac{3/2 + 4/3}{2} = 17/12 \text{ and } 2/a_3 = 24/17.$$

Continuing, the next average is

$$a_4 = \frac{17/12 + 24/17}{2} = \frac{577}{408}.$$

Our value, $\frac{577}{408} = 1.414215686\ldots$, is actually an overestimate of $\sqrt{2}$. And besides, we can't write the fraction $\frac{577}{408}$ as a terminating sexagesimal expansion. However, if we truncate it by continually subtracting negative powers of 60 from it, we do get

$$\frac{577}{408} = 1 + \frac{24}{60} + \frac{51}{3600} + \frac{10}{216,000} + \ldots$$

If we stop at this step without determining any remaining terms, we get the same exact Babylonian value indeed. This is admittedly a bit speculative but could well be the method used for this problem.

Here is another Babylonian problem dealing with depositing money with interest. How long would it take to double a bank deposit with 20% annual compounded interest? (Yes, their interest rates were astronomically high!)

Solution: Note that if it were simple interest, then the answer would simply be 5 years since an additional one-fifth of the principle would be added each year. But their answer was vvv; <<<< vvvvvvv < vvv <<= 3; 47, 13, 20 *years.*

The money had to be deposited with some sort of compound interest.

Here's an explanation. Deposit one unit of money.

One year later we have $6/5 = 1$; 12 units.
Two years later we have $(6/5)^2 = 36/25 = 1$; 26; 24 units.
Three years later we have $(6/5)^3 = 216/125 < 2$ units.
After four years we would have $(6/5)^4 = 1296/625 > 2$ units.

The correct answer is somewhere between 3 and 4 years. But they didn't just take the mean average. Instead, their value is a linear interpolation of the two values. Here

$$216/125 = 1080/625 < 2 = 1250/625 < 1296/625.$$

The difference between 2 and 216/125 is 170/625 while the difference between 1296/625 and 2 is just 46/625. The total difference between 216/125 and 1296/625 is 216/625. Thus, the value 2 is 170/216 of the way between the amounts at three years and four years respectively. Hence, their "solution" is given by adding 170/216 of a year to three years.

Thus, they arrive at the somewhat reasonable answer of $3 + \frac{170}{216}$ years, which when written sexagesimally becomes 3; 47, 13, 20.

In our next chapter, we will learn more about repeating decimals which will require us to fully understand geometric series. We'll take it slowly—one step at a time. But first, try your hand (and mind) on some of the following exercises.

Exercises

1. Determine the decimal expansions for the fractions 1/16, 1/40, 1/625.
2. Determine the decimal expansions for the fractions 3/16, 131/40, 9/125.

3. Determine the decimal fractions for 1/7, 2/7, 3/7, 4/7, 5/7, 6/7. Any interesting patterns? All have repeating decimal expansions of length 6. What is the sum of the first three digits with the next three digits for each of these fractions?

4. Determine the decimal fractions for 1/13, 2/13, 3/13, . . . , 12/13. Notice that these have decimal expansions of length 6 as well. Can you determine two related subclusters?

5. Determine the reduced common fractions represented by the decimals 0.12, 0.245, 6.0001.

6. Determine the common fraction represented by the decimal $0.121212\overline{12}.\dots$

7. Determine the common fraction represented by $0.1511\overline{1}.\dots$

8. Determine possible values of the three sexagesimal numbers

$$<<< v \qquad << vv, \qquad 0; <<, \qquad 0; << v <<$$

9. Determine the fractional value of the infinite sexagesimal expansion

$$0; < v < v < v\overline{< v}.\dots$$

10. What is the value of the fraction having ternary (base 3) expansion $0.0222\overline{2}\dots$?

11. Use our averaging algorithm to approximate $\sqrt{2}$ to a high degree of accuracy by starting with $a_1 = 7/5$ and calculating a_2 and a_3.

12. Verify that the sexagesimal 3; 47, 13, 20 is equal to $3\frac{170}{216}$.

13. After the French Revolution, the metric system was adopted. It was even suggested to divide a day into 10 equal hours, each delineated by 100 minutes of 100 seconds apiece. How would a second in this decimal system compare with our current second?

14. Middlebury/Williams Green Chicken Contest problem 1999: Three students are chosen randomly. Is it more likely that at least two were born on the same day of the week or that none of them were born on a weekend?

Chapter 2

Geometric Series, Powers of Three

> The happiness of life is made up of minute fractions—the little, soon forgotten charities of a kiss or a smile, a kind look or heartfelt compliment.
>
> **Samuel Taylor Coleridge (1772–1834),**
> **from "The Friend" (1828)**

In Chapter 1, we began by getting a solid grasp on fractions having finite decimal expansions. This included fractions like $1/2 = 0.5$, $1/4 = 0.25$, $1/5 = 0.2$, $1/32 = 0.03125$, and $1/500 = 0.002$. We didn't need to look far afield to see that not all fractions have such terminating decimals. To further develop our understanding of fractions, let's experiment with some denominators that are powers of 3. As they say, good things come in threes. We've already verified these interesting patterns:

$$1/3 = 0.33\bar{3}\ldots, \ 1/3^2 = 1/9 = 0.11\bar{1}\ldots, \text{ and } 1/3^3 = 1/27$$
$$= 0.037\overline{037}.\ldots$$

Next, let's tackle $1/81$. Since $81 = 3^4$, we must find the value of n for which a string of n 9's is divisible by 3^4. Unfortunately, this isn't so simple. It turns out the smallest string of 9's divisible by 3^4 is the number

$$10^9 - 1 = 999{,}999{,}999 = 3^4 \times 37 \times 333{,}667,$$

an integer made up of nine 9's. Thus, the result is that

$$1/81 = 37 \times 333{,}667/999{,}999{,}999.$$

Curiously, in this case, 37 × 333,667 = 12,345,679. Hence

$$1/81 = 0.0\overline{12345679}012345679...$$

with the first nine digits repeated forever. The decimal seems to follow the ordinary counting numbers before going just a bit off the tracks. What's happening here? To develop a fuller understanding, we will need to discuss infinite sums, a topic that has a long, arduous history.

A full investigation of the decimal expansion of 1/81 involves geometric series—in fact, a series of geometric series. We will take a deeper dive into infinite series later in the book (Chapter 14). Don't worry if this seems much less familiar. Together, we'll develop everything you need as we go.

The story of geometric series goes back at least as far as to Zeno of Elea, a Greek philosopher who lived in the fifth century BCE. Zeno was a student of Parmenides, the founder of the Eleatic school of philosophy competing with the Pythagoreans. The Pythagorean motto that "all is number" was challenged by the Eleatics who felt that time and motion were illusory. The paradoxes of Zeno that have come down to us (through the writings of Aristotle) attempt to challenge the very notions that time and space are either discrete or continuous in nature. Any natural assumption seems to end in paradox. In particular, the Achilles paradox argues against the assumption that time and space are continuous and thus can be subdivided ad infinitum.

Suppose we have a race between a very fleet-of-foot hare and a slow and plodding tortoise with the tortoise given a slight head start. By the time the hare reaches the starting position of the tortoise, the tortoise will have moved forward enough to maintain his lead. In the time that the hare moves to reach that point, the tortoise will have again moved forward slightly. This process will continue an infinite number of times, rendering it impossible for the hare to traverse the infinite number of steps necessary in a finite amount of time, and hence make it impossible to ever pass the tortoise. But of course, we know that the hare will, in fact, run pass the tortoise at some point.

According to Zeno, there is something amiss in our understanding of what seems like our common understanding of the continuity of space and time. It's certainly a subtle argument and seemed especially so to the ancient Greeks who only believed in "potential infinities" rather than in their actualization. In other words, they had little understanding of infinite processes or of our modern understanding of convergence versus divergence.

A similar thought experiment involves the inability to walk out of a room where the door is say 8 feet from you initially. First you must walk half the distance to the door (4 feet), then half again (2 feet), then half again (1 foot), then half again (1/2 foot), ad infinitum. You're never going to make it out of this room! Such infinite sums are now called geometric series.

In general, a geometric series is a series (an infinite sum) where each successive term is the same constant multiple of the previous term. For example,

$$8 + 4 + 2 + 1 + \tfrac{1}{2} + \ldots$$

is a geometric series with initial term 8 and each successive term one-half of the one before. The series

$$2 + \frac{2}{3} + \frac{2}{9} + \frac{2}{27} + \ldots$$

is another geometric series with initial term 2 and each successive term one-third of the preceding one. In general, we'll denote the first term by the letter a and the ratio of successive terms by r. Hence, all geometric series can be expressed as

$$a + ar + ar^2 + ar^3 + \ldots.$$

A series *converges* to a finite sum S if its sum gets arbitrarily nearer and nearer to S as ever more terms of the series are added. However, if the sum keeps surpassing all possible finite limits, then we say that the series *diverges*. It turns out that the two series above each converge to a finite number, specifically to the numbers 16 and 3, respectively. However, a series like

$$\frac{1}{2} + \frac{1}{2} + \frac{1}{2} + \frac{1}{2} + \ldots$$

with $a = \frac{1}{2}$ and $r = 1$ obviously will diverge since its sum will eventually exceed any finite value.

Convergence depends solely on the value of r rather than on the value of a. When the ratio r is small enough (less than 1 in absolute value), the geometric series will converge. The main result is the geometric series

$$a + ar + ar^2 + ar^3 + \ldots = a/(1-r) \text{ whenever } |r| < 1.$$

To see why this is so, let the sum

$$S = a + ar + ar^2 + ar^3 + \ldots.$$

Multiply both sides of the equation by r:

$$rS = ar + ar^2 + ar^3 + \ldots.$$

Subtracting the series for rS from that for S, all but one term remains. We get

$$S - rS = a.$$

Rewriting, we have

$$S(1-r) = a \text{ or } S = a/(1-r).$$

The proviso that $-1 < r < 1$ is needed so that the sum S adds up to a finite number rather than getting ever larger without bound. Thus, the geometric series *converges* to $a/(1-r)$ as long as r has absolute value strictly less than 1.

Here are two examples: First, if $a = 1$ and $r = \frac{1}{2}$, then the geometric series $1 + 1/2 + 1/4 + 1/8 + \ldots$ converges to 2. There's even an old joke dependent on this: An infinite number of mathematicians walk into a bar. The first mathematician tells the bartender, "I'll have a glass of your finest beer, my friend here will have half a glass, the next person half of half a glass, the next person . . . ." Rather than looking defeated, the bartender cuts in, "Yeah, okay I get it. Here's two glasses of beer for all of you!"

Another geometric series example is by letting $a = 1/3$ and $r = 1/5$, we obtain $1/3 + 1/15 + 1/75 + 1/375 + \ldots = (1/3)/(1 - 1/5) = 5/12$.

With our new perspective, let's take a look back at the decimal expansion for 1/9. From our previous discussion, we know that $1/9 = 0.11111\overline{1}$. Let's check using geometric series:

$$0.11111\overline{1} = 1/10 + 1/100 + 1/1,000 + 1/10,000 + 1/100,00$$
$$+ 1/1,000,000 + \ldots$$
$$= 1/10 + (1/10)^2 + (1/10)^3 + (1/10)^4 + (1/10)^5$$
$$+ (1/10)^6 + \ldots.$$

This is a geometric series with $a = 1/10$ and $r = 1/10$. Hence the series converges to $\frac{a}{1-r} = \frac{1/10}{1-1/10} = \frac{1}{9}$.

We now return to our discussion of the fraction 1/81. Consider adding together all the decimal fractions

$$0.01111\overline{1}\ldots + 0..0011111\overline{1}\ldots + 0.0001111\overline{1}\ldots + 0.000011111\overline{1}\ldots$$
$$+ \ldots \text{ ad infinitum.}$$

Each fraction is one-tenth the one preceding it. The first fraction is 1/10th of $0.11111\overline{1}\ldots.$

Hence, it is equal to $1/10 \times 1/9 = 1/90$. The next decimals are 1/900, 1/9000, 1/90,000, etc. Their sum also forms a geometric series having the sum

$$(1/90) + (1/900) + 1/9000) + \ldots \text{ with } a = 1/90 \text{ and } r = 1/10.$$

Lo and behold, its sum is

$$(1/90)/(1 - 1/10) = (1/10)(10/9) = 1/81.$$

If we write the sum vertically, we have that 1/81 is equal to

$$0.0111111111\bar{1}$$
$$+\, 0.0011111111\bar{1}$$
$$+\, 0.0001111111\bar{1}, \text{ etc.}$$

When we add down each column (tenth's place) beginning from the left, we get 0, then 1, then 2, then 3, etc. But when we add numbers longhand, we tend to start from the right side. The 1's gradually pile up, which results in carry overs. This is why the pattern of digits 0, 1, 2, 3, 4, etc. which seem to follow the counting (or natural) numbers doesn't go on forever. And besides, the number 0.012345678910111213 ... consisting of all the natural numbers in order has no repeating pattern of digits. So that number must be some other *irrational* number. Since 1/81 is a rational number, the pattern of consecutive natural numbers in its decimal expansion must begin to repeat at some point.

For interest's sake, here is another way to view the decimal expansion for 1/81. Notice that

$$\frac{1}{81} = \frac{1}{9} \cdot \frac{1}{9} = (0.1111\bar{1})(0.1111\bar{1})$$
$$= (0.1111\bar{1})(0.1 + 0.01 + 0.001 + 0.0001 + 0.00001 + \ldots)$$
$$= 0.01111\bar{1} + 0.001111\bar{1} + 0.0001111\bar{1} + \ldots,$$

which is identical to the series we derived for 1/81 earlier.

The next number to consider is $1/243 = 1/3^5$, one that fascinated the renowned physicist Richard Feynman (1918–88). In fact, 1/243 begins as 0.00411522633744855 ... and then (according to Feynman) soon goes a bit crazy before settling back down. A similar phenomenon is at play here as was the case with 1/81. The pattern 411, 522, 633, 744, 855, etc. can't go on forever. In fact,

$$1/243 = 0.0041152263374485596707818930\overline{0041152263374485596}$$

$$\overline{70781893}\ldots$$

with the first twenty-seven digits repeating ad infinitum. This can also be viewed as a more complicated geometric series with carry-overs interfering with the emerging pattern.

As we take larger powers of 3 in the denominator, the aesthetics of the fractions seem to diminish. In Chapter 3, we'll take a look at one of my favorite decimal expansions, namely that of 1/89. But for a bit of a change of pace, let's consider some more concrete examples involving thirds.

The German set theorist and mathematician Georg Cantor (1845–1918) considered the following set C (now called the Cantor set): Begin with the unit interval [0, 1] on the real number line and remove the open middle third (1/3, 2/3). What remains are two separate intervals [0, 1/3] U [2/3, 1]. Next, remove the middle thirds in each subinterval obtaining the set [0, 1/9] U [2/9, 1/3] U [2/3, 7/9] U [8/9, 1]. At this stage, four disjoint closed subintervals remain. Continue this center removal process ad infinitum. What remains is the Cantor set C. In an 1883 paper, Cantor developed many of its topological properties. For example, the set has measure zero although there are an uncountably infinite number of points left in C. (Actually, the first mention of this set was due to H. J. S. Smith in 1874.) Here we are simply interested in showing that the length of the sets removed add up to 1, confirming that what remains has measure zero.

For the second iteration, two sets each of length 1/9 were removed. For the third iteration, four sets of length 1/27 are removed, and so on (see Figure 2.1). The lengths removed are $\frac{1}{3}$, $2\left(\frac{1}{3}\right)^2$, $2^2\left(\frac{1}{3}\right)^3$, ..., $2^{n-1}\left(\frac{1}{3}\right)^n$, ... The sum of the lengths of all the removed subintervals forms the geometric series $\frac{1}{3} + \frac{2}{9} + \frac{4}{27} + \ldots$ with leading term $a = 1/3$ and ratio $r = 2/3$. Its sum is $\frac{a}{1-r} = 1$. Despite the fact we've removed intervals summing to the length of the original unit interval [0, 1],

Figure 2.1 First six iterations for the Cantor set. Reproduced from 127 "rect" (2007), public domain, via Wikimedia Commons.

the Cantor set C, far from being empty, still contains infinitely many points! This may seem completely paradoxical.

To gain a better understanding of the situation, we need to work in base 3. Just as all real numbers (points on a number line) have decimal expressions (base 10), they can be expressed in base 3 (ternary expressions) as well. The point 0 corresponds to $0.000\bar{0}$ in any base (including base 10 and base 3). The point 1 corresponds to either 1.0 or equivalently to $0.999\bar{9} = 9/10 + 9/100 + 9/1000 + \dots$ in base 10 and to $0.222\bar{2} = 2/3 + 2/9 + 2/27 + \dots$ in base 3. The midpoint ½ corresponds to 0.5 or $0.499\bar{9}$ in base 10 but $0.111\bar{1} = 1/3 + 1/9 + 1/27 + \dots$ in base 3. In forming the Cantor set, the first iteration removed all points with first ternary digit 1 while leaving those with first ternary digit 0 or 2. (Though it's a slight abuse of terminology, we'll use the phrase *ternary digit* to denote the numbers in the ternary expansion of a real number.) The second iteration removed all remaining points with second ternary digit 1 while leaving those with first and second ternary digits 0 or 2, and so on. The process continues ad infinitum, with the Cantor set itself containing those points between 0 and 1 with ternary expansion just containing 0's and 2's. Equivalently, the Cantor set C is the set of real numbers between 0 and 1 inclusive that do not require the number 1 in their ternary expansions. It's now clear that there are infinitely many elements of the Cantor set including such numbers (written in their ternary expansion) as $0, 0.0222\bar{2} = 1/3$, $0.2 = 2/3$, $0.02 = 2/9$, $0.22 = 8/9$, and $0.222\bar{2} = 1$. Though the set contains infinitely many points, it has measure 0 and hence contains no intervals within it. Technically, it's known as a *nowhere dense* set since its closure has empty interior.

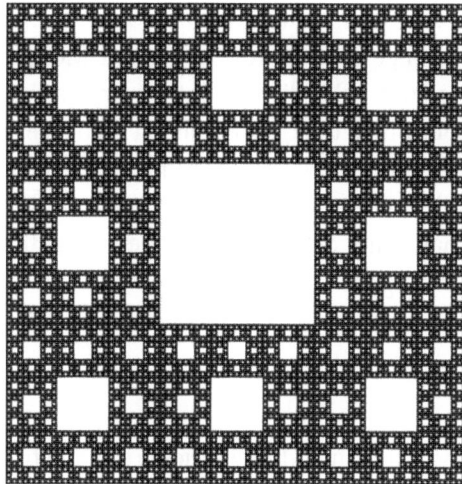

Figure 2.2 Sixth iteration of Sierpiński carpet. Reproduced from Johannes Rössel (2008), public domain, via Wikimedia Commons.

The Polish mathematician Wacław Sierpiński (1887–1969) developed a beautiful fractal-like two-dimensional analog of the Cantor set known as the Sierpiński carpet (1918). Sierpiński was a prolific researcher in many mathematical fields including set theory, number theory, function theory, and topology. He published 724 papers and fifty books along with founding and editing several pre-eminent mathematical journals including *Fundamenta Mathematicae.* As usual, though his name is associated with this construction, he himself gave full credit to his doctoral student Stefan Mazarkiewicz for having discovered it a year earlier. In any event, here is the construction.

Begin with a unit square and divide it into nine congruent squares (much like a tic-tac-toe board). Remove the central square. Eight congruent squares remain. Next, divide each of them into nine congruent squares and then remove the central square from each of them. Continue the process indefinitely. Figure 2.2 depicts the first few iterations.

The Sierpiński carpet is a rather sparse but connected set that includes the original square border. It may resemble a moth-infested macrame more so than it does a carpet. It has total area zero which you can confirm in the Exercises. And even the Sierpiński carpet itself admits several generalizations. One is a three-dimensional analog called the Menger sponge (1926) created by Karl Menger (1902–85). It begins with a unit cube divided into twenty-seven smaller congruent cubes from which the central cube in each face is removed along with the cube sitting smack dab in the very center. What is left are twenty smaller cubes. Now rinse and repeat . . . forever. The resulting figure has many of the same properties as the Sierpiński carpet. Lots to investigate!

Exercises

1. Determine whether the following infinite series are geometric series:
   (a) $1 + \frac{1}{2} + \frac{1}{3} + \frac{1}{4} + \frac{1}{5} + \frac{1}{6} + \ldots$
   (b) $2 + \frac{2}{5} + \frac{4}{25} + \frac{8}{125} + \frac{16}{625} + \frac{32}{3125} + \ldots$
   (c) $1 + \frac{1}{4} + \frac{1}{9} + \frac{1}{16} + \frac{1}{25} + \frac{1}{36} + \ldots$
   (d) $3 - 1 + \frac{1}{3} - \frac{1}{9} + \frac{1}{27} - \frac{1}{81} + \ldots$
   (e) $\pi + \pi^2 + \pi^3 + \pi^4 + \pi^5 + \pi^6 + \ldots .$

2. Determine whether the following geometric series converge. If so, find their sum.
   (a) $\frac{1}{5} + \frac{1}{25} + \frac{1}{125} + \frac{1}{625} + \ldots$
   (b) $\frac{1}{5} + \frac{6}{25} + \frac{36}{125} + \frac{216}{625} + \ldots$
   (c) $\frac{1}{2} + \frac{1}{8} + \frac{1}{32} + \frac{1}{128} + \ldots$
   (d) $\frac{1}{7} + \frac{2}{21} + \frac{4}{63} + \frac{8}{189} + \ldots$
   (e) $1 - \frac{1}{2} + \frac{1}{4} - \frac{1}{8} + \ldots .$

3. Write the following decimals as geometric series and then determine their common fraction:
   (a) $0.111111\bar{1}$
   (b) $0.121212121\bar{2}$
   (c) $2.343434\bar{34}$
   (d) $0.123123123123\overline{123}.$

4. Determine as much of the decimal expansion as you can for the fraction 100/729. Note that $729 = 3^6$. Any interesting patterns?

5. Use the fact that the Cantor set C is the set of all real numbers between 0 and 1 that can be written using just 0s and 2s in base 3 to exhibit five numbers that lie in the Cantor set and five numbers that do not.

6. (a) In the Sierpiński carpet, verify that the area of the first square removed is 1/9 and that eight squares of area 1/81 each is removed next.

   (b) Determine that the total area removed is $\frac{1}{9} + \frac{8}{9} \cdot \frac{1}{9} + \left(\frac{8}{9}\right)^2 \cdot \frac{1}{9} + \ldots$

   (c) Show that the above geometric series has sum 1 to conclude that the Sierpiński carpet has area zero.

7. The Greek mathematician Diophantus (3rd century CE) made sophisticated improvements to algebra and number theory in his collection *Arithmetica*. An epigram describing his life was published in a Greek anthology *c.*500 CE. It states, "He was a boy for one-sixth of his life, he married after a seventh part more, he grew a beard after another twelfth, and five years later granted him a son. Alas, the boy lived just half as long as his father. After consoling his grief through the science of numbers, four years later he ended his life." Accordingly, how long did Diophantus live?

Chapter 3
Fibonacci Numbers

Life is hardly more than a fraction of a second. Such a little
time to prepare oneself for eternity!

Paul Gauguin (1848–1903)

The Italian mathematics scholar Leonardo of Pisa (*c*.1180–1250),
better known as Fibonacci (son of Bonaccio) (Figure 3.1), lived at a
time when there was a great deal of cultural activity and commerce
among the port cities of the Mediterranean. As a young child, he
traveled extensively throughout Egypt, Syria, Greece, and Sicily with
his father, a merchant and customs manager of a large business firm
in Algeria. This brought the young Fibonacci in contact with Eastern
and Islamic mathematical practices which were far advanced from
those being practiced in the heart of Europe at the time.

Fibonacci's most influential book was his *Liber Abaci* (Book of
Computation), which was first printed in 1202. This book gives a
full explanation of how to do arithmetic with the Hindu-Arabic
numeral system, the system now used universally. Rather than being
innovative, Fibonacci drew upon centuries of Indian and Islamic
mathematical advances, especially the work of al-Khowarizmi (*c*.800
CE) dealing with square roots, cube roots, and both linear and
quadratic equations. Much of the work deals with commercial trans-
actions and monetary conversions. The word he used for zero was
"zephirum" which eventually led to the words "zero" and "cipher."
Of special interest to us, a horizontal bar was used for division,
much as we use today. The algebra itself was essentially *rhetorical* in
that it was a good mix of expansive exposition with a few equations
included, rather than being all symbols and equations. Here's one
problem from the book: A lion can eat a sheep in four hours, a
leopard can eat a sheep in five hours, and a bear can eat a sheep

Figure 3.1 Leonardo Fibonacci (1905), oil painting by unknown Medieval artist.

in six hours. How long would it take them to eat a sheep if they all ate together? Try your hand at it in the Exercises.

Fibonacci participated in a mathematical contest hosted by the Holy Roman Emperor Frederick II of the Norman Kingdom of Sicily. The questions were written by the king's clever assistant John of Palermo. All were difficult. Fibonacci was the only one able to solve all three problems. The first problem was to find a rational number t such that $t^2 + 5$ and $t^2 - 5$ are each squares of rational numbers. The final answer is that $t = 41/12$. The solution involved expert use of Fibonacci's two squares identity

$$\left(a^2 + b^2\right)\left(c^2 + d^2\right) = (ac + bd)^2 + (ad - bc)^2.$$

Note that the identity shows that the product of two integers, each expressible as the sum of two squares, is itself expressible as the sum of two squares.

The second problem was to find a solution to the cubic equation $x^3 + 2x^2 + 10x = 20$. At the time, no one knew how to solve a general cubic equation. For this problem, Fibonacci first showed that one could not construct a solution using Euclidean tools alone (straightedge and compass). Then he proceeded via some algebraic methods to get an excellent approximation. His answer was that x is approximately 1;22,7,42,33,4,40 written sexagesimally! Converted to its decimal expansion, we get 1.3688081075 which is accurate to nine decimal places.

The third problem is right up our alley. It involves fractions more directly and hence I've saved it as exercise 11 at the end of this chapter.

Fibonacci also wrote *Liber Quadratorum* with an explication of his two squares identity, a short treatise called *Flos* (flower), and *Practica Geometricae* (1220). The last work was largely based on Euclid's now lost *Division of Figures* and a Hebrew text by Abraham bar Hiyya, *Treatise on Mensuration*. The *Practica Geometricae* included the theorem that the medians of a triangle (meeting at the centroid) divide each other in the ratio of 2:1. It also included a three-dimensional extension of the Pythagorean Theorem for the main diagonal of a rectangular parallelepiped.

Next, let's look at one of my favorite decimal fractions, that of 1/89. We begin by displaying its decimal expansion:

$$1/89 = 0.\overline{01123595505617977528089887640449438202247191}\ldots$$

with a pattern that repeats of length 44. Admittedly, this takes more effort to work out. Lurking within this expansion are the Fibonacci numbers made famous by the rabbit problem included in the *Liber Abaci*, "How many pairs of rabbits can be produced from a single pair in a year if every month each pair begets a new pair which from the second month on becomes productive?" Keeping track of pairs, this leads to the

Fibonacci sequence $\{f_i\}$ = 1, 1, 2, 3, 5, 8, 13, 21, 34, ...

Figure 3.2 Golden ratio $\varphi = 1/a = a/b$ where $a + b = 1$.

More precisely, we define the Fibonacci sequence $\{f_i\}$ recursively by

$$f_1 = 1, f_2 = 1, \text{ and } f_{n+2} = f_n + f_{n+1} \text{ for } n \geq 1.$$

It is reasonable to also define $f_0 = 0$ which is consistent with our recursive definition since $f_2 = f_1 + f_0$.

You might find it instructive to write out the first dozen or so Fibonacci numbers. Next, you might wonder if there's some closed formula for the Fibonacci numbers. Indeed, there is! It was independently discovered by several mathematicians but is usually called Binet's Formula after his proof and publication in 1843. It involves a number well known throughout mathematical history, namely the *golden ratio* φ.

Given a unit line segment, there is a unique position where it can be divided in such a way that the ratio of the entire line segment to the larger piece is the same as the ratio of the larger piece to the smaller one. That ratio is often called the golden ratio and appears explicitly in ancient Greek mathematics and architecture (see Figure 3.2).

In Figure 3.2, a line segment of unit length is divided into a longer piece of length a on the left and a shorter piece of length b on the right. By design, $1/a = a/b$. But $b = 1 - a$. Hence, $1/a = a/(1 - a)$. Solving for a, we obtain $a^2 + a - 1 = 0$. Using the quadratic formula and noting that a is positive, we get that $a = \frac{-1+\sqrt{5}}{2}$. The golden ratio is the ratio $1/a$ (or equivalently a/b). Hence the golden ratio is $\varphi = \frac{1+\sqrt{5}}{2}$.

The astronomer Johann Kepler (1571–1630) aptly noted, "Geometry has two great treasures: one is the theorem of Pythagoras; the other, the division of a line into extreme and mean ratio. The first we may compare to a measure of gold; the second we may name a precious jewel."

Binet's formula: Let $\varphi = \frac{1+\sqrt{5}}{2}$ and let $\bar{\varphi} = \frac{1-\sqrt{5}}{2}$. Then $f_n = \frac{1}{\sqrt{5}}$ $(\varphi^n - \bar{\varphi}^n)$ for all $n \geq 1$.

At first sight, the formula looks crazy. After all, the Fibonacci numbers are all integers, and yet Binet's formula includes irrationals like $\sqrt{5}$. But please calculate f_1, f_2, and f_3 using the formula and note that there is a lot of cancellation. Quite miraculous!

How do we demonstrate that Binet's formula always works? We do it in two steps:

> Step 1: We will show that if x is any number for which $x^2 = x + 1$, then $x^n = f_n x + f_{n-1}$ for $n \geq 2$.
>
> Step 2: We will show that the appropriate values of x are in fact φ and $\bar{\varphi}$. Then we do a bit of arithmetic to clean things up.

There is no way to verify a formula for each value of n individually when there are infinitely many cases to check. Hence, to establish step 1, we use *induction*. Formally, induction is the mathematical principle that if a nonempty subset S of the positive integers contains the number 1 and contains the integer $n + 1$ whenever it contains the integer n, then S is itself the full set of positive integers. Informally, induction is the mathematical equivalent of setting up an endless row of dominoes and then knocking them all down by simply toppling the first domino.

Let's take a quick time out to make sure you're comfortable with mathematical induction by demonstrating that the sum of the first n odd positive integers is n^2. We begin by checking the veracity of our statement on the initial case (usually $n = 1$). Here we have that the sum of the first odd number is 1^2. Since $1 = 1^2$, the initial step checks out. Next, we establish the inductive step. Assume that the sum of the first n odd numbers is n^2 and then show that the proposition holds for the sum of the first $n + 1$ odd numbers. We assume that $1 + 3 + \ldots + (2n - 1) = n^2$ for some n. Next,

$$1 + 3 + \ldots + (2n - 1) + (2n + 1) = [1 + 3 + \ldots + (2n - 1)] + (2n + 1)$$
$$= n^2 + (2n + 1) = (n + 1)^2.$$

This is the formula we sought for the sum of the first $n + 1$ odd numbers. By induction, the formula holds for all $n \geq 1$.

Here's a relevant joke.

QUESTION: How does a mathematician induce good behavior in her children?

ANSWER: She says, "You know I've corrected you before. And if I've told you n times, I've told you $n + 1$ times."

Now we return to step 1 of our proof of Binet's formula. First, we'll check that the formula for x^n works for the initial case when $n = 2$. Then we'll show that if it's true for x^n for some n, it's also true for the next power of x, namely x^{n+1}. The formula will then hold for $n = 2$, then $n = 3$, then $n = 4$, etc.

When $n = 2$, we are given that $x^2 = x + 1$. But since $f_1 = 1$ and $f_2 = 1$, we verify immediately that $x^2 = f_1 x + f_2$. Next, let us assume that for some $n \geq 2$ that $x^n = f_n x + f_{n-1}$. In that case,

$$x^{n+1} = x^n(x) = (f_n x + f_{n-1}) x = f_n x^2 + f_{n-1} x.$$

But by hypothesis, $x^2 = x + 1$. Hence,

$$x^{n+1} = f_n x^2 + f_{n-1} x = f_n (x + 1) + f_{n-1} x = (f_n + f_{n-1}) x + f_n = f_{n+1} x + f_n,$$

since $f_n + f_{n-1} = f_{n+1}$. We've established step 1.

Using the quadratic formula, verify that the roots of the equation $x^2 = x + 1$ (or equivalently, $x^2 - x - 1 = 0$) are indeed φ and $\bar{\varphi}$ From step 1, we obtain

$$\varphi^n = f_n \varphi + f_{n-1}$$

and

$$\bar{\varphi}^n = f_n \bar{\varphi} + f_{n-1}.$$

This implies that

$$\varphi^n - \bar{\varphi}^n = f_n(\varphi - \bar{\varphi}).$$

However, $\varphi - \bar{\varphi} = \sqrt{5}$. Therefore, we have

$$f_n = \frac{1}{\sqrt{5}} (\varphi^n - \bar{\varphi}^n) \text{ for all } n \geq 1.$$

Since we have an exact formula for the nth Fibonacci number, we can get a good handle on its rate of growth. We can calculate some specific ratios $\frac{f_2}{f_1}, \frac{f_3}{f_2}, \frac{f_{11}}{f_{10}}, \frac{f_{101}}{f_{100}}$, ad infinitum. But what is the long-term ratio of each successive Fibonacci number to the one before? In more technical lingo, what is $\lim_{n \to \infty} \frac{f_{n+1}}{f_n}$? Here $\lim_{n \to \infty}$ stands for the "limit as n approaches infinity."

Some preliminary observations: On the one hand, $\varphi = \frac{1+\sqrt{5}}{2} = 1.618. \ldots > 1$. As we take higher powers of φ, the result gets ever larger without bound. On the other hand, $\bar{\varphi} = \frac{1-\sqrt{5}}{2} = -0.618\ldots$ and so $-1 < \bar{\varphi} < 0$. The point to notice is that $|\bar{\varphi}| < 1$. Hence, as we take higher powers of $\bar{\varphi}$, the result gets smaller and smaller in absolute value. i.e.,

$$\lim_{n \to \infty} \bar{\varphi}^n = 0.$$

To determine the long-term growth of the Fibonacci numbers, we substitute our expression for f_n from Binet's formula into the ratio $\frac{f_{n+1}}{f_n}$ and then take the limit as n goes to infinity:

$$\lim_{n \to \infty} \frac{f_{n+1}}{f_n} = \lim_{n \to \infty} \frac{\frac{1}{\sqrt{5}} \left(\varphi^{n+1} - \bar{\varphi}^{n+1} \right)}{\frac{1}{\sqrt{5}} \left(\varphi^n - \bar{\varphi}^n \right)}$$

$$= \lim_{n \to \infty} \frac{\varphi^n \left(\varphi - \bar{\varphi}^{n+1}/\varphi^n \right)}{\varphi^n \left(1 - \bar{\varphi}^n/\varphi^n \right)}$$

$$= \lim_{n \to \infty} \frac{\varphi - \bar{\varphi}^{n+1}/\varphi^n}{1 - \bar{\varphi}^n/\varphi^n}$$

$$= \frac{\varphi - 0}{1 - 0} = \varphi.$$

Thus, each Fibonacci number is approximately φ times as large as the one before with the approximation getting better as we proceed farther along the Fibonacci sequence.

There is a bit silly, but somewhat useful application of this. It turns out that 1 mile is approximately 1.609 kilometers. This ratio is pretty close to φ. I live in Vermont, USA. When I cross its northern border into Quebec, Canada, the road signs suddenly switch from mph (miles per hour) to kph (kilometers per hour). Knowing the Fibonacci sequence gives me a handy means to convert my driving speed fairly accurately. I move up one step to switch from mph to kph and down one step to switch from kph to mph. For example, the first speed limit sign I come across in Canada is 90 kph. Since 90 is very close to 89 and 89 is the Fibonacci number right after 55, I approximate 90 kph by 55 mph. Silly perhaps, but also useful for avoiding speeding tickets!

Returning to our fraction $\frac{1}{89}$, we begin by discussing the *generating function* for the Fibonacci numbers. It's defined by letting

$$F(x) = \sum_{n=1}^{\infty} f_n x^n = f_1 x + f_2 x^2 + f_3 x^3 + f_4 x^4 + \ldots$$

$$= 1x + 1x^2 + 2x^3 + 3x^4 + \ldots . \qquad (3.1)$$

Don't freak out. This is known as a power series since it contains an infinite sum of powers of x. We'll study power series more fully in Chapter 14. For now, just think of this sum as an endless polynomial in terms of x with the coefficients being the Fibonacci numbers. And for any chosen value of x, we just get a very long sum of *numbers* (some sort of infinite series). The capital sigma $\sum$ simply means "sum" and n is known as the "index." For example, $\sum_{n=1}^{4} n^2 = 1^2 + 2^2 + 3^2 + 4^2 = 30$.

Next, multiply both sides in formula (3.1) by x:

$$xF(x) = \sum_{n=1}^{\infty} f_n x^{n+1} = f_1 x^2 + f_2 x^3 + f_3 x^4 + f_4 x^5 + \ldots$$

$$= x^2 + x^3 + 2x^4 + 3x^5 + \ldots . \qquad (3.2)$$

One more step. Multiply formula (3.1) by x^2:

$$x^2 F(x) = \sum_{n=1}^{\infty} f_n x^{n+2} = f_1 x^3 + f_2 x^4 + f_3 x^5 + f_4 x^6 + \dots$$
$$= x^3 + x^4 + 2x^5 + 3x^6 + \dots. \qquad (3.3)$$

Thus,

$$\left(1 - x - x^2\right) F(x) = f_1 x + (f_2 - f_1) x^2 + (f_3 - f_2 - f_1) x^3$$
$$+ (f_4 - f_3 - f_2) x^4 + \dots,$$

which follows from subtracting formulas (3.2) and (3.3) from formula (3.1).

But we know from the recursive definition of Fibonnaci numbers that $f_{n+2} = f_n + f_{n+1}$ for $n \geq 1$. So all terms on the right after the first vanish. Therefore,

$$\left(1 - x - x^2\right) F(x) = x + 0x^2 + 0x^3 + 0x^4 + \dots = x.$$

Solving for the Fibonacci number generating function $F(x)$, we get the nice result

$$F(x) = \frac{x}{1 - x - x^2}. \qquad (3.4)$$

Letting $x = 1/10$ (assuming it's legitimate to do so),

$$F(1/10) = 0.1\,(f_1) + 0.01\,(f_2) + 0.001\,(f_3) + 0.0001\,(f_4) + \dots$$
$$= (1/10)/[1 - (1/10) - (1/100)] = (1/10)/(89/100) = 10/89.$$

Divide by 10 (accomplished by moving the expansion over to the right one decimal place) to obtain

$$1/89 = 1/100 + 1/1000 + 2/10{,}000 + 3/100{,}000 + 5/1{,}000{,}000$$
$$+ 8/10{,}000{,}000 + 13/100{,}000{,}000 + 21/1{,}000{,}000{,}000 + \dots.$$

Written vertically, it may be clearer as

$$\frac{1}{89} = 0.01$$

$$+ 0.001$$

$$+ 0.0002$$

$$+ 0.00003$$

$$+ 0.000005$$

$$+ 0.0000008$$

$$+ 0.00000013 + \ldots .$$

When we sum, there will be carry overs for all numerators consisting of more than one digit. But, amazingly, the decimal 1/89 essentially encapsulates all the Fibonnaci numbers within its decimal expansion.

If you want to explicitly exhibit more terms of the Fibonacci sequence including those with two digits, let $x = 1/100$ in formula (3.4). After simplifying and dividing by 100, you can display the fraction 1/9899. See for yourself!

In Chapter 6, we will learn a bit more about the period length of repeating decimals. What patterns are lurking there for us to discover? But first, in Chapters 4 and 5 we cover some simple basics about modular arithmetic.

Exercises

1. List the first fifteen Fibonacci numbers using the recursive definition for them. Are any squares?
2. Determine the first four Fibonacci numbers using Binet's formula.
3. Confirm how the constant φ appears in dividing a unit line segment into "extreme and mean" ratios.

4. Show that the sum of the first n Fibonacci numbers satisfies the following:

$$f_1 + f_2 + \ldots + f_n = f_{n+2} - 1 \text{ for all } n \geq 1.$$

5. Show that $f_2 + f_4 + \ldots + f_{2n} = f_{2n+1} - 1$ for all $n \geq 1$.
6. Show that $f_1^2 + f_2^2 + \ldots + f_n^2 = f_n f_{n+1}$ for all $n \geq 1$.
7. Show that $f_n^2 + f_{n+1}^2 = f_{2n+1}$ for all $n \geq 1$.
8. Establish Cassini's Identity (1680): $f_{n+1} f_{n-1} - f_n^2 = (-1)^n$ for all $n \geq 1$.
9. Determine the decimal expansion for $1/9899$.
10. A lion can eat a sheep in four hours, a leopard can eat a sheep in five hours, and a bear can eat a sheep in six hours. How long would it take them to eat a sheep if they all ate together?
11. Solve the third contest problem written by John of Palermo by finding the smallest solution possible: Three men possess a pile of money, their shares being 1/2, 1/3, and 1/6. Each man takes some money from the pile until nothing is left. The first man returns 1/2 of what he took, the second returns 1/3 of what he took, the third 1/6 of what he took. Then the total so returned is divided equally among the men. It is found that each then possesses exactly what he is entitled to. How much money was in the original pile and how much money did each man take from the pile?
12. (a) Express 73 as the sum of two squares.
    (b) Express 101 as the sum of two squares.
    (c) Given that $7{,}373 = 73 \times 101$, use Fibonacci's two squares identity to express 7,373 as the sum of two squares. Can you find a second representation of 7,373 as the sum of two squares?
13. The Lucas numbers (named after Edouard Lucas) are defined recursively by $L_0 = 2$, $L_1 = 1$, $L_2 = 3$, and $L_{n+2} = L_n + L_{n+1}$ for $n \geq 1$.
    (a) Show that $f_{n-1} + f_{n+1} = L_n$ for $n \geq 1$ where f_n is the nth Fibonacci number.
    (b) Show that $L_{n-1} + L_{n+1} = 5f_n$ for $n \geq 1$.

Chapter 4
Congruence Classes

A man is like a fraction whose numerator is what he is and whose denominator is what he thinks of himself. The larger the denominator, the smaller the fraction.

Leo Tolstoy (1828–1910)

At this juncture in our journey, it is convenient to introduce a couple of new ideas and some basic vocabulary. The following definition is due to the German mathematician, Carl Friedrich Gauss (1777–1855), perhaps the greatest mathematician of all time (see Figure 4.1).

Figure 4.1 Carl Friedrich Gauss, painted by Christian Albert Jensen in 1840.

Gauss was a child prodigy who at age three was able to correct a bookkeeping error made by his father. As a teenager, he proved that every integer can be expressed as the sum of three triangular numbers, constructed a regular 17-sided polygon using just a compass and straight-edge (considered impossible at the time since no ancient geometer ever attempted it), and conjectured the Prime Number Theorem, which accurately describes the distribution of primes a full century before anyone could prove it. During a long and illustrious career he made fundamental contributions to almost all of pure mathematics including multivariable calculus, real and complex analysis, differential geometry, probability, and especially algebra and number theory. He also made breakthroughs in applied mathematics and science in general. This work included astronomy (including predicting the location of the asteroid Ceres), geodesy, and electricity and magnetism (being a co-inventor of the electronic telegraph). Here we begin with a very useful concept he introduced in his work *Disquisitiones Arithmeticae.*

Definition: Let $n > 1$ be an integer. The integers a and b are said to be *congruent* to each other *modulo n* if n divides evenly into $a - b$. We denote this by $a \equiv b \pmod{n}$.

For example, $19 \equiv 5 \pmod 7$ since 7 divides evenly into $14 = 19 - 5$. We write this as $7|14$. A good way to think of this is that both numbers 5 and 19 have the same remainder when we divide them by 7. A congruence relation looks sort of like a regular equation, but with a triple of horizontal lines for the "equal" sign. This seems reasonable since 19 and 5 look and behave similarly relative to the number 7, though might not do so relative to a different modulus. Hence, although $19 \equiv 5 \pmod 7$, it's the case that 19 is *incongruent* to 5 modulo 3. The number 7 partitions all the integers—positive, negative, and zero—into seven groups (called *residue classes* modulo 7). Residue is just a fancy name for remainder. The classes are $\{\ldots, -7, 0, 7, 14, 21, \ldots\}$, $\{\ldots, -6, 1, 8, 15, 22, \ldots\}$, $\{\ldots, -5, 2, 9, 16, \ldots\}$, $\{\ldots, -4, 3, 10, 17, \ldots\}$, $\{\ldots, -3, 4, 11, 18, \ldots\}$, $\{\ldots, -2, 5, 12,$

19, . . .}, and {. . . , −1, 6, 13, 20, . . .}. Every integer resides in exactly one of these residue classes. Check your understanding by determining which residue class each of the numbers −30 and 100 reside modulo 7.

More generally, any positive integer $n > 1$ will partition the set of integers into n residue classes modulo n. A *complete set of residues modulo n* is a list of numbers with one representative chosen from each residue class. For example, {0, 1, 2, 3, 4, 5} forms a complete set of residues modulo 6 since every integer is congruent mod 6 to exactly one number in our set. Similarly, {6, 11, 16, 21, 26, 31} is another complete set of residues modulo 6. To cover our bases, we need six numbers, and each must represent a different residue class modulo 6. This is somewhat analogous to selecting a governor from each state to represent all the constituents in that state. Notice that the sets {2, 4, 6} and {1, 2, 3, 4, 5, 6, 7, 8} do not form complete residue sets modulo 6. The first set does not include all the residue classes modulo 6 while the second set includes redundancy (which is also not allowed).

Congruence modulo n forms what is known as an *equivalence relation*, meaning that it satisfies three basic properties referred to as *reflexivity*, *symmetry*, and *transitivity*. Just for the record, more specifically this is what we mean: If a, b, and c are integers and $n > 1$, then

(a) $a \equiv a \pmod{n}$ (reflexivity),
(b) If $a \equiv b \pmod{n}$, then $b \equiv a \pmod{n}$ (symmetry), and
(c) If $a \equiv b \pmod{n}$ and $b \equiv c \pmod{n}$, then $a \equiv c \pmod{n}$ (transitivity).

Equality itself is the paramount example of an equivalence relation of course. Many of the usual rules of arithmetic that we apply to equations can also be applied to congruence relations. Recall that you can add, subtract, multiply, or divide (by any number other than 0) on both sides of an equation without disturbing the equivalence. For example, since $5 = 3 + 2$, it follows that $10 \cdot 5 = 10(3 + 2)$. Next, we list some of the analogous results for congruences:

Result 1: Let $n > 1$ be a positive integer. Suppose $a \equiv b(\bmod n)$ and $c \equiv d(\bmod n)$. Then

    (a) $a + c \equiv b + d \,(\bmod n)$,
    (b) $a - c \equiv b - d \,(\bmod n)$, and
    (c) $ac \equiv bd \,(\bmod n)$.

Hence, we can add, subtract, and multiply expressions with a common congruence just as we do with normal equations. If our base n is 12, then this is often called "clock arithmetic." For example, three hours after 11:00 is 2:00 (at least in the United States).

I have been avoiding long technical arguments and difficult mathematical proofs, but in this case let's see how one might go about demonstrating result 1(a). Since $a \equiv b \,(\bmod n)$, it follows that $n|(a - b)$. Hence when we divide $a - b$ by n, we get an integer as quotient; call it k. So $nk = a - b$. Similarly, since $c \equiv d \,(\bmod n)$, there is an integer l for which $nl = c - d$. Adding these *equations* together, we get that

$$nk + nl = (a - b) + (c - d).$$

But the left-hand side of the equation is equivalent to $n(k + l)$ and the right-hand side can be rewritten as $(a + c) - (b + d)$. Since k and l are integers, so is $k + l$. Hence,

$$n \mid [(a + c) - (b + d)].$$

It follows that

$$a + c \equiv b + d \,(\bmod n),$$

which establishes result 1(a).

You may have noticed that, so far, we've avoided discussing division with congruences. Division requires greater care. For example, although it is true that

$$3 \cdot 5 = 15 \equiv 35 = 7 \cdot 5 \,(\bmod 10),$$

upon division by 5, it is not the case that 3 is congruent to 7 modulo 10. The problem is that the divisor 5 shares a factor in common with the modulus 10.

In what follows, we'll let gcd(a, b) denote the *greatest common divisor* of a and b. For example, gcd(6, 9) = 3 and the gcd(24, 40) = 8. There is a beautiful method called the Euclidean Algorithm due to Euclid (*c.*300 BCE) for finding the gcd of any two integers, but in most of our simple cases we can readily factor the numbers to find the largest integer dividing them both. For now, to not get sidetracked; factoring will be sufficient. But we'll dig a bit deeper into the Euclidean Algorithm shortly since it's extremely useful and ubiquitous in algebra and number theory.

We now state the general result for the division operation with congruence relations:

Result 2: Let d = gcd(a, n) and suppose that ax ≡ ay(mod n). Then x ≡ y(mod n/d).

Unlike the three previous operations, with division the modulus can change as well. In our previous example, $3 \times 5 \equiv 7 \times 5$ (mod 10) and gcd(5, 10) = 5. Hence, upon division by 5,

$$3 \equiv 7\,(\bmod\ 10/5)\ \text{or}\ 3 \equiv 7\,(\bmod\ 2).$$

Summarizing, adding, subtracting, and multiplying on both sides of a congruence relation has no effect on the modulus, but dividing might have some effect. If the divisor is relatively prime to the modulus, then there is no effect. But if the divisor has something in common with the modulus, we need to remove that factor in the modulus as well. It is somewhat like noting there is little danger and minimal long-term effects in removing someone's appendix since it's not a vital organ, but a surgeon must take much greater care with the removal of a kidney or a lung.

Let's combine our two results to solve the following for the variable x:

$$15x + 12 \equiv 6x + 9\,(\bmod\ 21).$$

We begin by subtracting 9 from both sides of our congruence relation,

$$15x + 3 \equiv 6x \,(\mathrm{mod}\ 21).$$

Now we divide both sides by the gcd(15, 6) = 3. We obtain

$$5x + 1 \equiv 2x \,(\mathrm{mod}\ 7).$$

Next subtract $2x$ from each side to get

$$3x + 1 \equiv 0 \,(\mathrm{mod}\ 7).$$

We can subtract 1 from each side to get $3x \equiv -1 \,(\mathrm{mod}\ 7)$. But $-1 \equiv 6 \,(\mathrm{mod}\ 7)$ since they both have the same remainder upon division by 7. We rewrite to get

$$3x \equiv 6 \,(\mathrm{mod}\ 7).$$

Now we divide by 3 to obtain

$$x \equiv 2 \,(\mathrm{mod}\ 7).$$

The modulus didn't change in this case since gcd(3, 7) = 1; i.e., the numbers 3 and 7 are relatively prime. Our final answer is correct, but since the original problem involved the modulus 21, we often rewrite our final answer to reflect that. (A question is best answered in the same language as it was asked.) The numbers x that are congruent to 2 modulo 7 are the numbers 2, 9, and 16 modulo 21 (if we list all the nonnegative solutions below 21). Hence our answer is

$$x \equiv 2, 9, \textit{ and } 16 \,(\mathrm{mod}\ 21).$$

Check to see that they are indeed all correct solutions to our original congruence.

At this point, some perspective about the mathematicians discussed in this book is in order. In various forms, mathematics has been studied for millennia throughout the globe. Its beauty and deep insights are appealing to all who are fortunate enough to have the leisure, support, and opportunities to study it. Many of the discoveries recounted here are so ancient, there is no way to know who was responsible for its inception. But in more modern times, a mathematical theorem can often be attributed to its discoverer. No doubt you will notice that until quite recently, most mathematicians were men. Unfortunately, for most of the world's history, women in particular have not been afforded anywhere near the support required. Too often they have been, in fact, explicitly barred from participating in the academic world. Even so, there are some brave and notable exceptions. One such person is Sophie Germain (see Figure 4.2).

Figure 4.2 Portrait of Sophie Germain.
Reproduced from Wikimedia Commons.

Sophie was born into a well-to-do Parisian family in 1776. She taught herself Latin and Greek, and then studied all the mathematics that she could in her father's extensive library. With little support from her family, who in fact tried everything to prevent it, she studied at night secretly in her cold bedroom by candlelight. When she got old enough, though women were barred from attending classes, she began getting course notes from a fellow student at the Ecole Polytechnique. She also submitted solutions using the pseudonym of a former student there, Monsieur LeBlanc, to homework exercises in a course taught by Joseph Louis Lagrange. Soon she did the same in other mathematics classes.

Eventually she read Gauss's *Disquisitiones Arithmeticae* and fell completely under the spell of number theory. By November of 1804, she began writing directly to Gauss about her discoveries but still referring to herself as Monsieur LeBlanc. In 1807 during the Napoleonic Wars, she used her family's political influence to ensure Gauss's personal safety by corresponding directly with General Pernety, who was in charge of the French forces occupying Gauss's hometown of Braunschweig. When General Pernety met Gauss and mentioned Sophie Germain by name, Gauss was confused. Within three months, Sophie Germain had to admit the truth of the matter. Gauss responded,

> How can I describe my astonishment and admiration on seeing my esteemed correspondent M. Le Blanc metamorphosed into this celebrated person . . . when a woman, because of her sex, our customs and prejudices, encounters infinitely more obstacles than men in familiarizing herself with [number theory's] knotty problems, yet overcomes these fetters and penetrates that which is most hidden, she doubtless has the noblest courage, extraordinary talent, and superior genius.

They corresponded a couple of times thereafter, but never met. However, in subsequent years, Germain continued her mathematical studies and made some wonderful discoveries. She corresponded with many of the greatest French scientists and mathematicians of the age including Legendre, Fourier, and Cauchy. She worked side by side with Poisson, though he published without acknowledging her influence or contributions. Even so, she was the first woman to receive a prize from the vaunted Paris Academy of Sciences in 1816 for her work on elasticity. It was her third attempt to solve a problem that had vexed her and all the other competitors for the prize, a true testament to her dedication and tenacity.

But she held the deepest affection for number theory. She made some very significant discoveries related to Fermat's Last Theorem, at that time the conjecture (now theorem) that there are no non-trivial solutions to the Diophantine equation $x^n + y^n = z^n$ for any integer $n \geq 3$. At the time, only a few small values of n had been established. Germain's theorem states that if p is a prime for which $2p + 1$ is also prime, then there is no solution in integers x, y, z for which $x^p + y^p = z^p$ where p does not divide xyz (the so-called first case of Fermat's Last Theorem). Subsequently, she was able to extend this somewhat to other primes p where $2np + 1$ was prime for various n. Due to this and other significant work, Gauss eventually recommended that the University of Gottingen bestow upon her an honorary doctoral degree. Unfortunately, she passed away in 1831 before such an occasion was possible. Still may her memory serve as inspiration for all women and men who follow in her footsteps.

Exercises

1. Determine which of the following sets form a complete residue system modulo 8.
   (a) $\{-3, -2, -1, 0, 1, 2, 3, 4\}$
   (b) $\{10, 20, 30, 40, 50, 60, 70, 80\}$
   (c) $\{0, 1, 2, 3, 4, 5, 6, 7\}$
   (d) $\{1, 2, 3, 4, 5, 6, 7, 8, 9, 10\}$
   (e) $\{1, 3, 5, 7\}$.
2. Solve the linear congruence $2x + 5 \equiv x - 9 \pmod{8}$.
3. (a) Solve the linear congruence $4x - 2 \equiv x + 4 \pmod{7}$.
   (b) Solve the linear congruence $4x - 2 \equiv x + 4 \pmod{9}$.

4. Verify that $x \equiv 3 \pmod 7$ and $x \equiv 4 \pmod 7$ give all solutions to the quadratic congruence $x^2 \equiv 2 \pmod 7$.

5. Find all solutions to $x^2 \equiv 1 \pmod 8$.

6. (a) Show that all squares are congruent to either 0 or 1 (modulo 4).

   (b) Show that no number in the sequence $3, 33, 333, 3333, \ldots$ is a perfect square.

   (c) Besides the numbers 1, 4, and 9, are any repeating digits a perfect square?

7. What is the remainder when a googol (10^{100}) is divided by 9? What about divided by 11?

8. Show that the product of any four consecutive numbers is divisible by 24.

9. How many consecutive zeros appear at the end of 1000! (1000 factorial)?

10. (One of my first arithmetic discoveries:) Show that n has the same remainder upon division by 11 as does the alternating sum of digits of n (beginning with the unit digits on the right).

11. Amy tells Andrew she's thinking of a number between 1 and 100 that is divisible by 3, and is congruent to 4 (mod 5) and congruent to 6 (mod 7). What is the number?

12. Henry tells Chris he's thinking of a number between 1 and 1000 that is congruent to 6 (mod 7), 1 (mod 11), and 12 (mod 13). What is the number?

13. Show that n and n^5 have the same final digit for any positive integer n.

14. Which of the properties reflexivity, symmetry, and transitivity are held by the following "relations"? (a) Being a first cousin, (b) being taller, (c) being at least as old, and (d) living in the same country.

15. Find all the Germain primes (primes for which p and $2p + 1$ are both prime) less than 100.

Chapter 5
Euclidean Algorithm

The only way I can distinguish proper from improper fractions is by their actions.

Ogden Nash (1902–71)

Recall that the greatest common divisor (gcd) of two positive integers is the largest integer that divides each of them. A straightforward method for finding the gcd has been known and utilized since ancient times. In fact, given integers a and b, the fact that the gcd(a, b) exists and is unique appears as Proposition 2 of Book VII in Euclid's *Elements* (*c*.300 BCE). The proof of this result involves an algorithm universally referred to as the Euclidean algorithm. It involves repeated division and determining remainders (i.e., residues). Rather than going through its proof initially, let's see an example which shows how it works.

Example of the Euclidean Algorithm: Find gcd(120, 654).

Solution: Begin by dividing the larger number by the smaller number and noting the remainder. This is known as the division algorithm. Alternatively, keep subtracting the smaller number from the larger as long as it's possible to avoid a negative remainder (or keep adding the smaller number to itself as long as the sum stays below the larger number). Here we have

$$654 = 5 \times 120 + 54.$$

Next, we continue to apply the division algorithm to 120 and 54:

$$120 = 2 \times 54 + 12.$$

We repeat with the last two remainders:

$$54 = 4 \times 12 + 6.$$

On our next step we get $12 = 2 \times 6$ with a zero remainder. The last *nonzero remainder* is the gcd we seek. So $\gcd(120, 654) = 6$.

The Euclidean algorithm is very efficient and has a useful bonus. Having completed the process, it can be reversed to write the gcd of two numbers as a linear combination of them. In this case,

$$6 = 54 - 4 \times 12.$$

But

$$12 = 120 - 2 \times 54.$$

Substituting the last equation in the former, we get

$$6 = 54 - 4(120 - 2 \times 54).$$

Simplifying,

$$6 = -4 \times 120 + 9 \times 54$$

But

$$54 = 654 - 5 \times 120,$$

hence

$$6 = -4(120) + 9(654 - 5 \times 120) = 9 \times 654 - 49 \times 120.$$

We have found a linear combination of 654 and 120 that equals the greatest common divisor of them. In general, writing the greatest common divisor of two integers as a linear combination of them is referred to as *Bezout's identity*.

One nice consequence is that if two integers are relatively prime (have no common factors), then there is a linear combination of them equaling 1. That is, if $\gcd(a, b) = 1$, then there exists x and y for which $ax + by = 1$. The converse is also true: if there exists an x and y for which $ax + by = 1$, then a and b are relatively prime. For if $\gcd(a, b) = d$ and $d > 1$, then $d | ax$ and $d \mid by$ for any x and y. But then $d | (ax + by)$ which contradicts our hypothesis that $ax + by = 1$.

This has direct application in solving congruence relations as well. Recall that it is legitimate to add, subtract, and multiply on both sides of a congruence relation. However, division is trickier. For example, consider the linear congruence relation

$$8x + 7 \equiv 3x + 5 \pmod{11}. \tag{5.1}$$

To solve, we begin by subtracting $3x$ from both sides of the congruence

$$5x + 7 \equiv 5 \pmod{11}.$$

Next, we subtract 7 from both sides

$$5x \equiv -2 \pmod{11}. \tag{5.2}$$

We can't divide both sides by 5 since $-2/5$ isn't even an integer. In Chapter 4, we would find an integer divisible by 5 that is congruent to -2 modulo 11. This idea works but seems a bit ad hoc not knowing how to go about finding such a number. In this case, $-2 \equiv 75 \pmod{11}$ and 75 is divisible by 5. So we can rewrite the congruence as

$$5x \equiv 75 \pmod{11}$$

then divide by 5

$$x \equiv 15 \pmod{11}$$

and finally simplify by replacing 15 with a smaller residue obtaining $x \equiv 4 \pmod{11}$.

The Euclidean algorithm provides an alternative method for solving congruence relations without the need for any guess work. Look back at equation (5.2). Since the numbers 5 and 11 are relatively prime, Bezout's identity guarantees that there are numbers m and n for which $5m + 11n = 1$. Equivalently, $5m \equiv 1(\mathrm{mod}\ 11)$. The number m is known as the *arithmetic inverse* of 5 modulo 11. That is, multiplying 5 by m has the same effect as if we were dividing by 5. The arithmetic inverse of a number a is often denoted by a^*. For example, we see that $5^* \equiv 9\ (\mathrm{mod}\ 11)$ which we can find by utilizing the Euclidean algorithm followed by applying Bezout's identity. Check that $5 \cdot 9 = 45 \equiv 1\,(\mathrm{mod}\ 11)$.

Returning to equation (5.2), multiply both sides by the arithmetic inverse of 5 (mod 11)

$$9(5x) \equiv 9(-2)\,(\mathrm{mod}\ 11).$$

But, by design, $9 \cdot 5 = 45 \equiv 1(\mathrm{mod}\ 11)$ and $9(-2) = -18 \equiv 4(\mathrm{mod}\ 11)$. We obtain $x \equiv 4(\mathrm{mod}\ 11)$ as before.

You might want to try your hand with the following short exercises. For each pair of numbers, find their greatest common divisor, and then express their gcd as a linear combination of the original pair: (36, 111), (16, 206), (89, 144).

For greater completeness, let's actually prove that the Euclidean algorithm always works. Mathematicians always want to know why something is true and whether there are any conditions or limitations involved. (If you work through the proof, the earlier example should help as a concrete example of the steps involved.)

We begin by carefully defining a greatest common divisor. Given two positive integers a and b, the integer d is the *greatest common divisor* of a and b if

(a) $d > 0$ (for consistency and for the sake of uniqueness, we insist that d be positive),

(b) $d \mid a$ and $d \mid b$ (we say d *divides* a and d *divides* b), and

(c) If $f \mid a$ and $f \mid b$, then $f \mid d$ (this makes d the *greatest* of the common divisors).

Proposition: *Let a and b be positive integers. Then gcd(a, b) exists and is unique.*

Proof (Euclidean Algorithm): We divide a by b and obtain a quotient q_1 and remainder r_1. That is, apply the division algorithm to a and b. In particular, we have

$$a = q_1 b + r_1, \text{ where } 0 \leq r_1 < b.$$

We repeatedly apply the division algorithm as long as we keep getting positive remainders:

$$b = q_2 + r_2, \text{ where } 0 \leq r_2 < r_1.$$

$$r_1 = q_3 r_2 + r_3, \text{ where } 0 \leq r_3 < r_2.$$

$$r_{n-3} = q_{n-1} r_{n-2} + r_{n-1}, \text{ where } 0 \leq r_{n-1} < r_{n-2}.$$

$$r_{n-2} = q_n r_{n-1}, \text{ where } r_n = 0.$$

Here r_n is defined as the first zero remainder. This process must eventually terminate in n steps for some $n \geq 1$, since the remainders are strictly decreasing nonnegative integers. In fact, clearly n is at most the minimum of either a or b.

We claim that r_{n-1} is the greatest common divisor of a and b. To verify this, we must check the three conditions in the definition of gcd.

(a) By definition, $r_{n-i} > 0$.

(b) From our last equation, $r_{n-1} | r_{n-2}$. But then r_{n-1} divides both terms on the right in the penultimate equation. Hence $r_{n-1} | r_{n-3}$. Similarly, r_{n-1} divides $r_{n-4}, \ldots, r_1$, and so on. Thus, $r_{n-1} | b$ and $r_{n-1} | a$.

(c) If $f | a$ and $f | b$, then we have that $f | r_1$ since $r_1 = a - q_1 b$. But $r_2 = b - q_2 r_1$, so $f | r_2$. Similarly, f divides $r_3, \ldots, r_{n-2}$. Hence $f | r_{n-1}$.

To complete our proof, we need to show uniqueness, namely that there aren't additional numbers that satisfy conditions (a), (b), and (c). Let d be a gcd(a, b) satisfying the three conditions of the definition. Since r_{n-1} is a greatest common divisor, we have that $d|r_{n-1}$ and also that $r_{n-1}|d$. It follows that $d = r_{n-1}$ or $d = -r_{n-1}$. But d and r_{n-1} are both positive integers and so $d = r_{n-1}$. Hence, the greatest common divisor of a and b exists and is unique.

The Euclidean algorithm has other benefits as well. To recount one, recall the least common multiple (lcm) of two numbers is the least integer that is divisible by each of them. For example, lcm(30, 50) = 150 since 150 is the least integer divisible by both 30 and 50. We can check this by looking at the prime factorizations of 30 = 2 × 3 × 5 and 50 = 2 × 5^2 and then choose the fewest prime factors necessary to find a common divisor. In this case we need a 2, a 3, and two 5's. So lcm(30, 50) = 2 × 3 × 5^2 = 150. But what if the prime factorizations of the numbers weren't so apparent? Here we make use of the observation that the product of two given numbers, say a and b, will be a common multiple (though not necessarily the least of them). The number ab can then be divided by the gcd(a, b) and we still have an integer that is divisible by both a and b. This is so since all the factors common to both a and b appear twice in the product ab. Hence we can remove those factors (once) and still have a number divisible by each of a and b. However, the removal of any additional factors will result in a smaller number that will no longer be divisible by at least one of a or b. Hence, we get the nice result

$$\text{lcm}(a, b) = \frac{ab}{\gcd(a, b)}.$$

Hence the Euclidean algorithm is equally adept at finding lcm's as it is finding gcd's.

One More Example: Find the lcm(1260, 5805).

Solution: Apply the Euclidean algorithm. 5805 = 4 × 1260 + 765. Next, 1260 = 1 × 765 + 495. But 765 = 1 × 495 + 270, 495 = 1 × 270 + 225, 270 = 1 × 225 + 45, and 225 = 5 × 45. Thus, gcd(1260, 5805) = 45. It follows that lcm(1260, 5805) = 1260 × 5805/45 = 162,540.

Exercises

1. Determine (a) gcd(36, 111), (b) gcd(16, 206), (c) gcd(89, 144).

2. Express the greatest common divisors of the above pairs as linear combinations of them.

3. Find (a) lcm(36, 111), (b) lcm(16, 206), (c) lcm(89, 144).

4. (a) Show that if $gcd(m, n) = 1$ and d is any integer, then there exist integers x and y for which $mx + ny = d$. Hence, linear combinations of relatively prime integers can be found to sum to any integer whatsoever.

   (b) Find x and y for which $37x + 49y = 20$. (There are infinitely many correct answers.)

5. Show that if $gcd(a, b) = 1$, then in fact there are infinitely many pairs of integers x and y for which $ax + by = 1$.

6. A used bookstore sells all paperbacks for $3 apiece and all hardbacks for $8 apiece. Describe what can be purchased for precisely $100.

7. Find the arithmetic inverse of 17 (modulo 103) and then solve $17x \equiv 4 (\mathrm{mod}\ 103)$.

8. Find the arithmetic inverse of 71 (modulo 109) and then solve $71x \equiv 3 (\mathrm{mod}\ 109)$.

9. Let S be any set of $n + 1$ integers chosen from the set $\{1, 2, \ldots, 2n\}$. Show that S must contain two relatively prime integers.

10. (a) Find a positive integer at most 105 that is $\equiv 0 (\mathrm{mod}\ 3)$, $\equiv 2 (\mathrm{mod}\ 5)$, and $\equiv 3 (\mathrm{mod}\ 7)$.

    (b) Find a positive integer at most 1001 that is $\equiv 2 (\mathrm{mod}\ 7)$, $\equiv 1 (\mathrm{mod}\ 11)$, and $\equiv 9 (\mathrm{mod}\ 13)$.

    An ancient theorem known as the Chinese remainder theorem guarantees the existence of such solutions.

Chapter 6

Euler Phi Function and the Period Length of Repeating Decimals

Wherever there is number, there is beauty.

Proclus (412–85)

Recall that prime numbers are those positive integers $p > 1$ only divisible by 1 and p itself. Let us display some examples of their reciprocals. We determine that $1/7 = 0.142857\overline{142857}...$, $1/11 = 0.09\overline{09}...$, $1/13 = 0.076923\overline{076923}...$, and $1/7 = 0.05882352941176$ $470588235294117647....$ Each has a nonterminating but repeating decimal expansion. The fraction $1/7$ has a repeating pattern of length six, $1/11$ a pattern of length two, $1/13$ of length six, and $1/17$ of length sixteen. Is there any rhyme or reason to all of this? Is there a relationship between the prime p and the period length of $1/p$? And what about $2/p$ or $3/p$? And what about $1/n$ if n isn't a prime? These are all great questions which mathematicians have partially answered quite nicely. Perhaps you want to make some conjectures on your own at this point. Somewhat surprisingly, to this day no one has a totally complete answer to some of these questions. But we will make some good progress in this chapter.

We begin with a definition.

Definition: For each natural number $n \geq 1$, the Euler phi function $\varphi(n)$ is the number of positive integers less than or equal to n that are relatively prime to n.

For example, the positive integers less than or equal to 6 that are relatively prime to 6 are 1 and 5 (since 2, 3, 4, and 6 each share some common factor with 6). Hence $\varphi(6) = 2$ since there were just two such numbers. The set of numbers $\{1, 2, 3, 4, 5, 6\}$ form a *complete set of residues* modulo 6 since every integer is congruent to exactly one of them mod 6; whereas the set {1, 5} form a *reduced set of residues* modulo 6. The integers less than or equal to 7 that are relatively prime to 7 are 1, 2, 3, 4, 5, and 6 (since 7 is prime). Thus, {1, 2, 3, 4, 5, 6} form a reduced set of residues modulo 7 and $\varphi(7) = 6$. We define $\varphi(1) = 1$, which is consistent with the wording of our definition.

This function was carefully studied by the great Swiss mathematician Leonhard Euler (1707–83) and hence has been named after him. He was one of the most prolific mathematicians of all time despite the fact that he was visually impaired for most of his adult life. In fact, he was completely blind towards the end of his career but showed no diminution in his deep insights, computational abilities, and awe-inspiring productivity. Euler worked in or created almost every area of modern mathematics including number theory, algebra, calculus, analysis, differential equations, combinatorics, and graph theory. His collected works include over a thousand letters of correspondence plus 866 scientific and mathematical papers and reports, many of them thick volumes full of original mathematics.

What patterns are waiting to be discovered here? Let us attempt to find a formula for $\varphi(n)$ that works for all n. Note that $\varphi(p) = p - 1$ for any *prime* number p since all positive integers less than p are relatively prime to it. Hence $\varphi(23) = 22$, for example, and $\varphi(89) = 88$. What about integers that are powers of a prime such as $9 = 3^2$, $125 = 5^3$, or $2401 = 7^4$? Let's work more generally to determine $\varphi(p^n)$ for any prime p and integer n. The number p^n is only divisible by a single prime, namely p itself. So it shares that factor with p, $2p$, $3p$, etc.—all numbers that are multiples of p. Every pth number is divisible by p. (For example, for $p = 7$, the numbers 7, 14, 21, etc. are all divisible by 7 while all other numbers not multiples of 7 are relatively prime to 7.) So the fraction $1/p$ of numbers from 1 to p^n are not relatively prime to p^n while all the other numbers are relatively prime to p^n. For example, every third number in the list 1, 2, 3, 4, 5, 6, 7, 8, 9 shares the factor 3 with the number 9 (namely, 3, 6, and 9). The other numbers (1, 2, 4, 5, 7, 8) do not. Hence, 1/3 of the numbers from 1 to 9 share a common factor with 9 (i.e., a nontrivial factor larger than 1) while 2/3 of the numbers from 1 to 9 do not. In this case $\varphi(9) = (2/3)(9) = (1 - 1/3)(9) = 6$. One-third of the numbers from

1 to 9 are divisible by 3 while all the rest or $1 - 1/3 = 2/3$ of them are relatively prime to 9. More generally, $\varphi(p^n) = (1 - 1/p)p^n$ since the proportion of integers relatively prime to p^n (i.e., those numbers not divisible by p) is $1 - 1/p$ of all the numbers from 1 to p^n. We can rewrite this as

$$\varphi(p^n) = (1 - 1/p)p^n = p^n - p^{n-1} = (p - 1)p^{n-1}.$$

Let's highlight this result:

$$\varphi(p^n) = (p - 1)p^{n-1} \tag{6.1}$$

Of course, not all numbers are simply powers of a single prime number. Let's gain some insight by studying a simple example, say the number $15 = 3 \times 5$. It's important to note that the primes 3 and 5 are relatively prime to each other. To find $\varphi(15)$ we must determine the number of integers less than 15 that are relatively prime to 15. The numbers relatively prime to 15 are those that are neither multiples of 3 nor multiples of 5. Let us make a short chart of the numbers from 1 to 15 with their remainders when we divide by each of 3 and 5. Please recall that when any integer m is divided by another positive integer n, there will be a quotient q and a remainder r where $0 \le r < n$. This is the basis for the *division algorithm*.

Number n	Remainder upon division by 3	Remainder upon division by 5
1	1	1
2	2	2
3	0	3
4	1	4
5	2	0
6	0	1
7	1	2
8	2	3
9	0	4
10	1	0
11	2	1
12	0	2
13	1	3
14	2	4
15	0	0

The numbers relatively prime to 15 are those that have nonzero numbers in *both* columns displaying the remainder upon division by 3 and the remainder upon division by 5. Hence from our list of numbers from 1 to 15, we must eliminate those multiples of 3 and those of 5. Alternatively, the numbers relatively prime to 15 are those that are simultaneously relatively prime to 3 and to 5. Since 1/3 of the numbers are divisible by 3, $1 - 1/3 = 2/3$ of the numbers are relatively prime to 3. Similarly, 1/5 of the numbers are divisible by 5, and hence $1 - 1/5$ of the numbers are relatively prime to 5. Furthermore, other than $15 = 3 \times 5$ itself, no positive integer less than 15 is divisible by both 3 and 5 because 3 and 5 are relatively prime. So the total number relatively prime to 15 are

$$(1 - 1/3)(1 - 1/5)(15) = (2/3)(4/5)(15) = 8.$$

Reordering, we can rewrite this as

$$\varphi(15) = (2/3)(3)(4/5)(5) = \varphi(3)\varphi(5).$$

To determine $\varphi(15)$, we just needed to multiply $\varphi(3)$ by $\varphi(5)$.

What we observed above was that the phi function behaves *multiplicatively*. There was nothing special about the prime numbers 3 and 5. We could make the same argument with 7 and 17 or any other two primes. In fact, we could extend this notion to a product of three or more primes just as easily or even to products of distinct prime powers. The key is that the factors are relatively prime to one another.

The function φ being multiplicative means that if m and n are relatively prime, then $\varphi(mn) = \varphi(m)\varphi(n)$.

By the Fundamental Theorem of Arithmetic, every positive integer has a unique prime factorization. We now invoke formula (6.1). Although we haven't given a fully rigorous proof of the following, I hope that I've conveyed a sense of why the following seems entirely reasonable:

If $n = p_1^{a_1} p_2^{a_2} \dots . p_r^{a_r}$, then $\varphi(n) = (p_1 - 1) p_1^{a_1 - 1} (p_2 - 1) p_2^{a_2 - 1} \dots$
$$(p_r - 1) p_r^{a_r - 1} \tag{6.2}$$

Try some examples utilizing formula (6.2). See if you can determine $\varphi(21)$, $\varphi(27)$, $\varphi(105)$, $\varphi(1000)$. An important first step is to determine the prime factorization of the argument in each case. Here we get $21 = 3 \times 7$, $27 = 3^3$, $105 = 3 \times 5 \times 7$, and $1000 = 2^3 \times 5^3$. Using formula (6.2), we readily surmise that $\varphi(21) = 2 \times 6 = 12$, $\varphi(27) = 2 \times 3^2 = 18$, $\varphi(105) = 2 \times 4 \times 6 = 48$, and $\varphi(1000) = 1 \times 2^2 \times 4 \times 5^2 = 400$. We've turned what seemed like a long, difficult process into a fairly quick and easy computation.

We now turn our attention to some renowned results from number theory which will prove to be extremely useful. We begin with an example as usual. Consider the list of numbers 1, 2, 3, 4, 5, 6, which are the numbers less than 7 that are relatively prime to 7. Pick any number that is relatively prime to 7—say 3 (though we could choose any such number even those larger than 7). Multiply the numbers in our list by our chosen number 3. We get 3, 6, 9, 12, 15, 18. If we divide each one by 7 and note the remainder, our new list becomes 3, 6, 2, 5, 1, 4. (For example, $4 \times 3 = 12$ and $12 = 1 \times 7 + 5$, so we note the remainder 5.) Notice that our new set contains the exact same numbers as our original set, just in a jumbled order. In the language from Chapter 4, both contain the same list of residues modulo 7. Part of the reason is that the product of two numbers relatively prime to 7 will itself be relatively prime to 7. (This is a basic result from number theory called Euclid's lemma.) The other key ingredient is that $3a$ and $3b$ will have different remainders upon division by 7 as long as a and b are different relative to 7.

Let's try a different example to double-check our understanding. Consider the list of numbers less than 15 that are relatively prime to it: 1, 2, 4, 7, 8, 11, 13, 14. (We know that $\varphi(15) = 8$ by way of a double check.) Pick some number relatively prime to 15—say 4. Now multiply the members of our list by 4: 4, 8, 16, 28, 32, 44, 52, 56. Next, find their remainders upon division by 15. (Technically, we are reducing *modulo* 15.) We obtain 4, 8, 1, 13, 2, 14, 7, 11. Again we have the same list as the one we began with, just in a different order. Next, we work more generally.

Consider any number n and another integer a that is relatively prime to n. There are $\varphi(n)$ integers less than n that are relatively prime to n. (Recall, these numbers form a *reduced residue system modulo n*. Residue means remainder.) Let's list those residues as we did previously:

$$r_1, r_2, \ldots, r_{\varphi(n)}.$$

Of course, there are $\varphi(n)$ of them all told. Now form the list

$$ar_1, ar_2, \ldots, ar_{\varphi(n)}$$

and note again that it must be simply a reordering of our original list once we reduce modulo n (i.e., by replacing each number by its remainder upon division by n). Hence, the product of numbers $r_1 r_2 \ldots r_{\varphi(n)}$ and the product of the numbers $(ar_1)(ar_2) \ldots \left(ar_{\varphi(n)}\right)$ are identical (modulo n). We write this as

$$r_1 r_2 \ldots r_{\varphi(n)} \equiv (ar_1)(ar_2) \ldots \left(ar_{\varphi(n)}\right)(\bmod\, n).$$

This just says that the two sides of our "equation" are equal with respect to n. They have the same remainder upon division by n, but maybe not upon division by some other number. Both sides of our congruence equation have factors $r_1, r_2, \ldots, r_{\varphi(n)}$. Since each of them is relatively prime to n, so is their product. Hence, gcd $\left(r_1 r_2 \ldots \ldots r_{\varphi(n)}, n\right) = 1$. We can divide through by all of them to obtain 1 on the left-hand side and $\varphi(n)$ factors of a on the right-hand side. By Result 2 of Chapter 4, the modulus doesn't change when we divide in this instance. The miraculous result is stated below:

Euler–Fermat Theorem: *If a and n are relatively prime, then* $a^{\varphi(n)} \equiv 1\,(mod\, n)$.

This theorem is a celebrated number-theoretic result based on a key result due to the French mathematician Pierre de Fermat (1601–65) and a subsequent generalization and proof by Leonhard Euler. The earlier special case, known as Fermat's little theorem, states that if p is an odd prime and p doesn't divide a, then $a^{p-1} \equiv 1 \pmod{p}$.

Of course, this follows from the Euler–Fermat theorem since $\varphi(p) = p - 1$.

Let's look back at some of our examples. Let $n = 7$ and $a = 3$. Here $\varphi(7) = 6$. The Euler–Fermat theorem asserts that $3^6 \equiv 1 \pmod 7$. By way of verification, $3^6 = 729 = 7 \times 104 + 1$. This checks since we get a remainder or residue of 1. Next, let $n = 15$ and $a = 4$. We know that $\varphi(15) = 8$. The Euler–Fermat theorem says that $4^8 \equiv 1 \pmod{15}$. To check, $4^8 = 65,536 = 15 \times 4369 + 1$.

Now we can make use of a lot of our preliminaries by letting $a = 10$. Given an integer a with $\gcd(a, 10) = 1$, the Euler–Fermat theorem guarantees that $10^{\varphi(a)} \equiv 1 \pmod n$. Hence, n divides the number $10^{\varphi(a)} - 1$, which is the number consisting of $\varphi(a)$ 9's. This relates directly to the repeating decimal expansions we studied in Chapter 1. Given any integer a relatively prime to 10, the fraction $1/a$ will have an immediate repeating decimal expansion that repeats after every $\varphi(a)$ digits. So it must have a decimal expansion of period $\varphi(a)$ or a number that divides evenly into $\varphi(a)$ in order to repeat every $\varphi(a)$ digits. Very cool!

Let's look back at some of our examples. The number $1/7 = 0.142857\overline{142857} \ldots$ had period 6 and $\varphi(7) = 6$. The number $1/17 = 0.05882352941176470\overline{588235294117647} \ldots$ has period 16 and $\varphi(17) = 16$. The number $1/13 = 0.076923\overline{076923} \ldots$ has period 6 and $6|12$ where $\varphi(13) = 12$. The number $1/89$ had a repeating pattern consisting of 44 digits while $44|88$ and $\varphi(89) = 88$. The pattern length of $1/n$ is always a divisor of $\varphi(n)$. This is welcome news and a great help in determining and checking our results.

One deep conjecture (not quite completely proved) is that there are infinitely many primes p for which the period length of $1/p$ is $\varphi(p) = p - 1$ and nothing shorter (a proper divisor of $p - 1$). It is widely expected that for infinitely many primes p, the period length of $1/p$ is as long as is possible. This is actually part of a more general conjecture known as Artin's primitive root conjecture. For example, long before the invention of electronic computers, the English amateur mathematician William Shanks (1812–82) calculated the full 17,388-digit period for the fraction $1/17,389$. Why did he choose this particular prime? Perhaps it's because 17,389 is the 2,000th prime!

If you want to do a bit or your own investigating, compare the decimals for the various sevenths, namely 1/7, 2/7, 3/7, 4/7, 5/7, and 6/7. Note how the digits slide around. Next check to see what happens with all the thirteenths. Here you should see two separate groups, each with their own digit-sliding behavior. What about seventeenths? Please check it out.

What's going on? Let's look again at the division involved in the fraction 1/7. Ignoring the decimal point, we begin by dividing 7 into 10, getting quotient of 1 and remainder of 3. Next, we divide 7 into 30, getting a quotient of 4 and remainder of 2. But we can think of this second division as simply the first division in determining the decimal expansion of 3/7. The decimal expansion for 3/7 is one step behind that of 1/7. Continuing, we can verify that $1/7 = 0.\overline{142857}$ while $3/7 = 0.\overline{428571}$. In fact, since 1/7 has period length 6, each of the remainders 1, 2, 3, 4, 5, and 6 must occur in our calculation of 1/7. Hence, each of 1/7, 2/7, 3/7, 4/7, 5/7, and 6/7 have the same repeating decimal expansion, only that they have different starting points. The pattern 142,857 just slides around a bit. And of course, this isn't something special about the prime 7. The same phenomenon must hold for any prime p for which the decimal expansion of $1/p$ has maximum period length of $p - 1$. For these primes, the fractions $2/p$, $3/p, \ldots, (p - 1)/p$ will have the same digit pattern as $1/p$ but with different starting positions. Thus, the story of the fractions n/p for all $n \geq 1$ is actually told by the fraction $1/p$. The primes p for which $1/p$ has shorter period lengths are also interesting, but more intricate. Enjoy investigating the related patterns for n/p for various n with any of your favorite primes!

Exercises

1. Determine $\varphi(n)$ for the following values of n:
   (a) $n = 50$, (b) $n = 91$, (c) $n = 105$, (d) $n = 2027$, (e) $n = 1,000,000,000$.

2. Determine $\varphi(30)$ by listing all the numbers less than 30 that are relatively prime to 30. Note that your list contains all prime numbers. (In fact, 30 is the largest number having this property.)

3. Explain why each of the following forms a complete set of residues modulo 7:
   (a) $\{-3, -2, -1, 0, 1, 2, 3\}$, (b) $\{2, 4, 6, 8, 10, 12, 14\}$, (c) $\{0, 8, 16, 24, 32, 40, 48\}$.

4. Explain why each of the following does not form a complete set of residues modulo 7:
   (a) $\{1, 2, 3, 4, 5, 6, 7, 8\}$, (b) $\{1, 2, 3, 4, 5\}$, (c) $\{1, 2, 3, 10, 11, 12, 13\}$.

5. Explain why each of the following sets form a reduced set of residues modulo 7:
   (a) $\{1, 2, 3, 4, 5, 6\}$, (b) $\{-3, -2, -1, 1, 2, 3\}$, (c) $\{71, 72, 73, 74, 75, 76\}$.

6. Use the Euler–Fermat theorem to show that if p is a prime with $p = 3$ or $p > 5$, then p divides an integer consisting solely of a string of 1s. In fact, show that p will divide an infinite number of such strings.

7. (a) Calculate the decimal fraction for $1/13$.
   (b) Compare $1/13$ with $n/13$ for values $n = 2, 3, \ldots, 12$.
   (c) Add the first three digits of the decimal fraction for $1/13$ to the next three digits. Compare with $n/13$ for various values of n.

8. (a) Calculate the decimal fraction for $1/19$.
   (b) Compare $1/19$ with $n/19$ for values $n = 2, 3, \ldots, 18$.
   (c) Add the first nine digits in the decimal fraction for $1/19$ to the next nine digits. Compare with $n/19$ for various values of n.

9. Show that n^5 has the same last digit as n for any integer n.

10. What is the last digit of $2^{1000000}$?

11. (a) Verify that $\sum_{d|n} \varphi(d) = n$ for $n = 8$, $n = 11$, $n = 12$, and $n = 20$.

(b) Prove that for all $n \geq 1, \sum_{d|n} \varphi(d) = n$.

12. For which primes $p < 100$ does the fraction $1/p$ have maximum repetition length of $p - 1$?

13. Investigate the decimal expansions for $n/13$ for $n = 1, \ldots, 12$.

14. Investigate the decimal expansions for $n/37$ for $n = 1, \ldots, 36$.

Chapter 7

A Very Brief Introduction to Matrices

> Mathematics knows no race or geographic boundaries.
> **David Hilbert (1862–1943)**

Tabulating numbers in a rectangular array and operating on them collectively in some fashion dates back to ancient Chinese mathematicians. However, not until recent times has the concept of a matrix been formally defined and developed. James Joseph Sylvester (1814–97) coined the term in 1850 to mean an "oblong arrangement of terms" and he and Arthur Cayley derived many of its key properties in *Memoir of the Theory of Matrices* published in 1858. Matrices are now ubiquitous in the natural and social sciences, especially with the emergence of big data. So what exactly are we dealing with?

A *matrix* is simply a rectangular array of numbers such as $\begin{bmatrix} 1 & 3 & 2 \end{bmatrix}$ or $\begin{bmatrix} 4 & -1 \\ 3 & 7 \end{bmatrix}$ or $\begin{bmatrix} 1 & 0 & 1.5 \\ -0.12 & 17 & 9.4 \end{bmatrix}$. The *size* of the matrix is described by its number of rows by number of columns. Rows go across horizontally from left to right while columns go down vertically from top to bottom. The previous matrices are of size 1×3, 2×2, and 2×3 respectively. A *square matrix* has the same number of rows as columns.

We'll be especially interested in 2×2 matrices. If two matrices have the same size, we can add or subtract them component-wise. For example,

$$\begin{bmatrix} 1 & 2 \\ 3 & 4 \end{bmatrix} + \begin{bmatrix} 5 & 6 \\ 7 & 8 \end{bmatrix} = \begin{bmatrix} 6 & 8 \\ 10 & 12 \end{bmatrix},$$

since $1 + 5 = 6$, $2 + 6 = 8$, $3 + 7 = 10$, and $4 + 8 = 12$.

The 2 × 2 *zero matrix* is the matrix $O = \begin{bmatrix} 0 & 0 \\ 0 & 0 \end{bmatrix}$, having the useful property that $A + O = O + A = A$ for any 2 × 2 matrix A. Hence, O is the additive identity among 2 × 2 matrices analogous to the number 0 in ordinary arithmetic.

Multiplication of a matrix by a scalar (meaning any old number) is also carried out component-wise. For example,

$$5 \begin{bmatrix} 2 & 1 \\ 4 & 6 \end{bmatrix} = \begin{bmatrix} 10 & 5 \\ 20 & 30 \end{bmatrix},$$

where we simply multiply each component of our matrix by the scalar 5.

Every matrix has an additive inverse as well. The additive inverse of A is $-1A = -A$. The sum of a matrix plus its additive inverse results in a zero matrix. For example, the additive inverse of $\begin{bmatrix} 2 & 1 \\ 4 & 6 \end{bmatrix}$ is $\begin{bmatrix} -2 & -1 \\ -4 & -6 \end{bmatrix}$ and

$$\begin{bmatrix} 2 & 1 \\ 4 & 6 \end{bmatrix} + \begin{bmatrix} -2 & -1 \\ -4 & -6 \end{bmatrix} = \begin{bmatrix} 0 & 0 \\ 0 & 0 \end{bmatrix}.$$

Multiplication of two matrices is defined in a bit more intricate fashion. The product of two 2 × 2 matrices is defined as follows: Let $A = \begin{bmatrix} a & b \\ c & d \end{bmatrix}$ and $B = \begin{bmatrix} e & f \\ g & h \end{bmatrix}$. Then

$$AB = \begin{bmatrix} a & b \\ c & d \end{bmatrix} \begin{bmatrix} e & f \\ g & h \end{bmatrix} = \begin{bmatrix} ae + bg & af + bh \\ ce + dg & cf + dh \end{bmatrix}.$$

To determine the number in the ith row and jth column of AB (where both i and j equal either 1 or 2), we multiply elements of the ith row of A by the corresponding element of the jth column of B and add as we go along. Move from left to right along the ith row of A while moving from top to bottom along the jth column of B.

For example,

$$\begin{bmatrix} 1 & 2 \\ 3 & 4 \end{bmatrix} \begin{bmatrix} 5 & 6 \\ 7 & 8 \end{bmatrix} = \begin{bmatrix} 19 & 22 \\ 43 & 50 \end{bmatrix},$$

since $1 \times 5 + 2 \times 7 = 19$, $1 \times 6 + 2 \times 8 = 22$, $3 \times 5 + 4 \times 7 = 43$, and lastly $3 \times 6 + 4 \times 8 = 50$. You might find it of interest to note that $\begin{bmatrix} 5 & 6 \\ 7 & 8 \end{bmatrix} \begin{bmatrix} 1 & 2 \\ 3 & 4 \end{bmatrix}$ is a different 2×2 matrix. In general, AB ≠ BA. So matrix multiplication isn't *commutative*. Luckily, matrix multiplication is still *associative*. It follows that there is no ambiguity between A(BC) or (AB)C. Hence parentheses are unnecessary when we restrict ourselves to matrix multiplication. That is, ABC = A(BC) = (AB)C.

There is a special 2×2 matrix called the 2×2 *identity* matrix. It's $I = \begin{bmatrix} 1 & 0 \\ 0 & 1 \end{bmatrix}$. Note that if you take any 2×2 matrix A and multiply by I, the product is just A again. So A I = A and I A = A. Verify this with a matrix A of your choice. The matrix I acts just like the number 1 which is the multiplicative identity in ordinary arithmetic.

In ordinary arithmetic, most numbers have multiplicative inverses—namely the number which when multiplied by it results in the product 1. For example, the inverse of 5 is 1/5 and the inverse of $-1/2$ is -2. We write $5^{-1} = 1/5$ and $(-1/2)^{-1} = -2$. All real numbers r except zero have a multiplicative inverse, namely $1/r$. Similarly, many 2×2 matrices A have multiplicative inverses as well. If it exists, the multiplicative inverse of A, denoted A^{-1}, is the 2×2 matrix for which $AA^{-1} = A^{-1}A = I$. Some matrices, such as the zero matrix, do not have inverses. But there are also other matrices lacking inverses. It turns out that the inverse of $A = \begin{bmatrix} a & b \\ c & d \end{bmatrix}$ exists as long as $ad - bc \neq 0$. The expression $ad - bc$ is known as the *determinant* of the 2×2 matrix A. We denote this by

$$\det A = \det \begin{bmatrix} a & b \\ c & d \end{bmatrix} = ad - bc.$$

By the way, Gauss was the one who coined the term "determinant." The word matrix actually derives from a Latin word meaning womb or uterus. Apparently, the matrix determines or gives birth to its determinant.

Here is the main result:

$$\text{If } A = \begin{bmatrix} a & b \\ c & d \end{bmatrix} \text{ and } \det A = ad - bc \neq 0,$$

then A is invertible and

$$A^{-1} = \frac{1}{ad - bc} \begin{bmatrix} d & -b \\ -c & a \end{bmatrix}. \tag{7.1}$$

Matrices that have inverses are called *invertible*, while those not having inverses are called *singular*. We better take a look at an example.

Suppose we wish to find the inverse of the matrix $A = \begin{bmatrix} 1 & 2 \\ 3 & 4 \end{bmatrix}$. In this case, the determinant $ad - bc = 1 \times 4 - 2 \times 3 = -2$. Since $-2 \neq 0$, our matrix A has an inverse. In this case,

$$A^{-1} = \frac{1}{-2} \begin{bmatrix} 4 & -2 \\ -3 & 1 \end{bmatrix} = \begin{bmatrix} -2 & 1 \\ 3/2 & -1/2 \end{bmatrix}$$

Please check by verifying that

$$AA^{-1} = A^{-1}A = \begin{bmatrix} 1 & 0 \\ 0 & 1 \end{bmatrix} = I.$$

Now a little bit about matrix equations: A linear equation in two variables is an algebraic equation of the form $ax + by = c$ where a, b, and c are given and x and y are variables. Often we want to solve a system of linear equations. For instance, we might want to solve for x and y given that

$$3x + y = 11$$

$$2x + 3y = 12.$$

Our linear system of equations can be rewritten as a matrix equation

$$AX = B \tag{7.2}$$

Where $A = \begin{bmatrix} 3 & 1 \\ 2 & 3 \end{bmatrix}$, $X = \begin{bmatrix} x \\ y \end{bmatrix}$, and $B = \begin{bmatrix} 11 \\ 12 \end{bmatrix}$. To solve this system, we multiply A^{-1} on the left of both sides of equation (7.2). (Since matrix multiplication isn't commutative, we need to keep track on which side we multiply.) This results in $A^{-1}AX = A^{-1}B$, which can then be simplified to

$$X = A^{-1}B. \tag{7.3}$$

In this example, the determinant of A is $3 \times 3 - 1 \times 2 = 7 \neq 0$. Hence by formula (7.1),

$$A^{-1} = \frac{1}{7}\begin{bmatrix} 3 & -1 \\ -2 & 3 \end{bmatrix}$$

and thus

$$X = \begin{bmatrix} x \\ y \end{bmatrix} = \frac{1}{7}\begin{bmatrix} 3 & -1 \\ -2 & 3 \end{bmatrix}\begin{bmatrix} 11 \\ 12 \end{bmatrix} = \frac{1}{7}\begin{bmatrix} 21 \\ 14 \end{bmatrix} = \begin{bmatrix} 3 \\ 2 \end{bmatrix}.$$

Check that $x = 3$, $y = 2$ solves both of the original equations.

Exercises

Let $A = \begin{bmatrix} 1 & 2 \\ -1 & 0 \end{bmatrix}$, $B = \begin{bmatrix} 3 & 1 \\ 6 & 2 \end{bmatrix}$, $C = \begin{bmatrix} 1 & 0 & 3 \\ 2 & -2 & 1 \end{bmatrix}$.

1. Determine which of the following operations exist. If so, calculate the result.
   (a) A + B, (b) A + C, (c) AB, (d) AC, (e) CA, (f) (A–B)C.
2. Determine which of the matrices A, B, C have inverses. If so, find its inverse.
3. Solve for X: AX = B.

4. Find the determinants of the following 2×2 matrices. If the matrix has an inverse, find it.

$$A = \begin{bmatrix} 1 & 2 \\ 3 & 5 \end{bmatrix}, B = \begin{bmatrix} 2 & 3 \\ 5 & 4 \end{bmatrix}, C = \begin{bmatrix} 10 & 4 \\ 20 & 8 \end{bmatrix}, D = \begin{bmatrix} 1 & 0 \\ 0 & 1 \end{bmatrix}.$$

5. The matrix

$$I = \begin{bmatrix} 1 & 0 & 0 \\ 0 & 1 & 0 \\ 0 & 0 & 1 \end{bmatrix}$$

is the 3×3 identity matrix. Verify that if

$$A = \begin{bmatrix} 1 & 0 & 1 \\ 1 & 1 & 1 \\ 0 & -1 & 1 \end{bmatrix},$$

then

$$A^{-1} = \begin{bmatrix} 2 & -1 & -1 \\ -1 & 1 & 0 \\ -1 & 1 & 1 \end{bmatrix}.$$

Take a linear algebra class to learn how to *find* inverses of square invertible matrices of all sizes.

6. Verify that det AB = det $A \cdot$ det B for any choice of 2×2 matrices A and B. This helps explain why the definition of matrix multiplication is a natural one.

Chapter 8
Farey Fractions

> We are all functioning at a small fraction of our capacity.
> **Winston Churchill (1874–1965)**

In this chapter, we'll take a break from all the complications involved in adding, multiplying, and dividing fractions in the usual way. We are going to define a new and much simpler operation defined on special subsets of rational numbers called Farey fractions. Let's begin with a definition and a bunch of straight-forward examples. Recall a bit about the greatest common divisor. Check that $\gcd(14, 26) = 2$ and $\gcd(15, 50) = 5$. If $\gcd(a, b) = 1$, then we say that a and b are *relatively prime*. The numbers 10 and 21 are relatively prime, but 7 and 21 are not.

Definition 1: Let F_n denote the set of rational numbers a/b with $0 \leq a \leq b \leq n$ with $\gcd(a, b) = 1$ arranged in increasing order. Then the sequence F_n is the sequence of *Farey fractions of order n*.

Here are the Farey fractions of orders 1 through 6:

F_1: 0/1, 1/1
F_2: 0/1, 1/2, 1/1
F_3: 0/1, 1/3, 1/2, 2/3, 1/1
F_4: 0/1, 1/4, 1/3, 1/2, 2/3, 3/4, 1/1
F_5: 0/1, 1/5, 1/4, 1/3, 2/5, 1/2, 3/5, 2/3, 3/4, 4/5, 1/1
F_6: 0/1, 1/6, 1/5, 1/4, 1/3, 2/5, 1/2, 3/5, 2/3, 3/4, 4/5, 5/6, 1/1

Note that if a/b is a member of F_n, then it's a member of all succeeding F_m with $m \geq n$. Each new sequence F_n simply adds the reduced fractions between 0 and 1 having denominator n to the previous sequence $F_n - 1$. Each step of the way $\varphi(n)$ additional fractions are appended. Please add a line to our chart by determining F_7.

The Farey fractions are named after the British geologist and surveyor John Farey. Though he amassed a sizeable collection of rocks and minerals, he was not a particularly distinguished geologist. However, he did publish his thoughts and opinions on a variety of topics including music, comets, carriage wheels, and decimal coinage. In 1816, he discovered (or more likely re-discovered) and published some interesting properties of what we now call Farey fractions in the popular English periodical *Philosophical Magazine*. However, he did not furnish any justifications or mathematical proofs of his assertions. They were supplied that very same year by one of Europe's greatest mathematicians, Augustus-Louis Cauchy (1789–1857). Cauchy named the fractions after Farey, but in fact some other individuals had made similar observations previously, including Charles Haros in 1802. Such independent discoveries are not at all uncommon in mathematics.

At one time in Europe, the units for weights and measures depended on which country you lived in. In fact, it often varied from town to town and city to city. After the French Revolution, the government worked to establish a uniform metric system based squarely on base 10. To this end, Haros was employed to make several tables, including one that contained the decimal expansions of all reduced fractions between 0 and 1 with numerator less than or equal to 100. There are 3003 such fractions in all. To make sure that he didn't miss any fractions, Haros used the concept of the *mediant* of fractions (which we define shortly). In fact, he acknowledged that his algorithm was actually due to a Nicolas Chuquet, who utilized it the mid-seventeenth century. Starting with just 0/1 and 1/1, repeated application of the mediant gave Haros a full list of all such reduced fractions. When the mediant created a denominator greater than 100, he simply ignored the result.

Fifteen years later, Henry Goodwyn took on the mammoth task of extending Haros' table to include all reduced fractions with denominators up to $2^{10} = 1024$. His method of discovery was not described. Though he never quite completed his task (which would have included 318,963 entries), a partial list was circulated in 1816 under the title, "The First Centenary of Series of Concise and Useful Tables of all the Complete Decimal Quotients." This was the table seen by John Farey that led him to his discoveries.

The main result that interested all these folks was the following:

Proposition 1: *If $a/b < f/g < c/d$ are three consecutive terms in F_n for any n, then*

$$\frac{f}{g} = \frac{a+c}{b+d}.$$

Verify this relation on several of your own examples for various fractions and different Farey sequences. For example, In F_5, consider the three consecutive terms 3/5, 2/3, and 3/4. Check that

$$\frac{2}{3} = \frac{3+3}{5+4}.$$

Establishing the veracity of this general result is a bit involved. Its proof actually depends directly on another interesting observation, namely

Proposition 2: *If a/b and c/d are successive terms in F_n for some n, then $ad - bc = -1$.*

It is instructive to at least verify Proposition 2 on several of your own examples. Another way to state Proposition 2 is that if a/b and c/d are successive terms in F_n, then

$$\det \begin{bmatrix} a & c \\ b & d \end{bmatrix} = -1.$$

A complete proof involves quite a bit of detail which is not especially illuminating in this setting. Hence, I will not include it here. However, let's see how the truth of Proposition 2 implies that of Proposition 1: Let $a/b < f/g < c/d$ be three consecutive terms in F_n.

Assuming Proposition 2, we have $ag - bf = -1$ and $fd - gc = -1$. So $ag - bf = fd - gc$ and $ag + gc = fd + bf$. Rewriting, $(a + c)g = f(b + d)$. Thus,

$$\frac{f}{g} = \frac{a + c}{b + d}.$$

Our discussion above leads us to define a new and refreshingly simple operation on fractions.

Definition 2: The *mediant* of the fractions a/b and c/d is the fraction $\frac{a+c}{b+d}$. We write the mediant operation as $a/b \oplus c/d$.

For example,

$$1/5 \oplus 1/3 = \frac{1 + 1}{3 + 5} = 2/8 = 1/4.$$

Similarly,

$$1/8 \oplus 2/3 = 3/11.$$

The mediant of two different fractions will always be a fraction lying between them. This follows from noting that if $a/b < c/d$, then $ad < bc$ and so $ab + ad < ab + bc$. But this can be rewritten as $a(b + d) < b(a + c)$ which implies that $a/b < (a + c)/(b + d)$. Similarly, $a/b < c/d$ implies that $ad < bc$ which implies that $ad + cd < bc + cd$ or $d(a + c) < c(b + d)$. This then implies that $(a + c)/(b + d) < c/d$.

There is a nice way to picture all of this and see (visually) that the mediant lies strictly between the fractions that created it. The slope of the first triangle is a/b and the slope of the second triangle is c/d. The slope of the line stretching across both of them is the mediant which clearly lies between the slopes of the two triangles (Figure 8.1). Again,

$$\frac{a}{b} < \frac{a + c}{b + d} < \frac{c}{d}.$$

Furthermore, it's worth noting that the denominator of the mediant of two consecutive fractions in F_n has denominator greater

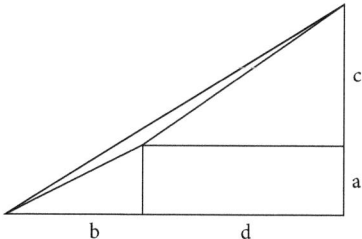

Figure 8.1 Slopes of the three hypotenuses: $a/b < (a+c)/(b+d) < c/d$.

than n. This is true since a/b and c/d are consecutive terms in F_n. Hence their mediant is not a member of F_n and thus $b + d > n$.

There are several important mathematical applications that utilize Farey fractions. One key application involves the approximation of real numbers by rationals. Here is the result:

Proposition 3. *If x is any real number and n a positive integer, then there is a reduced fraction p/q such that $0 < q \leq n$ and*

$$|x - p/q| < \frac{1}{q(n+1)}.$$

Basically, Proposition 3 says that given any real number x (say, $\sqrt{2}$ or $5 + \sqrt[3]{7}$ or π), we can find a rational number with not too big a denominator that approximates it fairly well. Of course, the larger the denominator, the better the approximation. Here is a proof of Proposition 3.

Proof: Without loss of generality, assume that $0 \leq x < 1$. Let $a/b \leq x < c/d$ where a/b and c/d are successive terms in F_n. Then either

(i) x lies in the interval

$$\left(\frac{a}{b}, \frac{a+c}{b+d} \right) \ or$$

(ii) x lies in the interval

$$\left(\frac{a+c}{b+d}, \frac{c}{d} \right).$$

On the one hand,

$$\frac{a+c}{b+d} - \frac{a}{b} = \frac{bc-ad}{b(b+d)} = \frac{1}{b(b+d)}$$

by Proposition 2. But $b + d \geq n + 1$. So in case (i), let $p = a$ and $q = b$. Then

$$\left| x - \frac{p}{q} \right| < \frac{1}{b(b+d)} < \frac{1}{q(n+1)}.$$

On the other hand,

$$\frac{c}{d} - \frac{a+c}{b+d} = \frac{bc-ad}{d(b+d)}.$$

So in case (ii), let $p = c$ and $q = d$ and

$$\left| x - \frac{p}{q} \right| < \frac{1}{d(b+d)} < \frac{1}{q(n+1)}.$$

Finally, if it's the case that $m \leq x < m + 1$ for some integer m, then $x = m + x_0$ where $0 \leq x < 1$. From the proof above, there is a reduced fraction $\frac{p_0}{q_0}$ such that

$$\left| x_0 - \frac{p_0}{q_0} \right| < \frac{1}{q(n+1)}.$$

Since $\gcd(p_0, q_0) = 1$, it follows that $\gcd(mq_0 + p_0, q_0) = 1$. So let $p = mq_0 + p_0$ and $q = q_0$ and the proof is complete.

For example, suppose we wish to find a fraction $\frac{p}{q}$ such that

$$\left| e - \frac{p}{q} \right| < \frac{1}{8q},$$

where the denominator q must satisfy $1 \leq q \leq 7$. (i.e., with a fairly small denominator). Here $e = 2.71828\ldots$ is the base of the natural logarithm function. Let $x = e$. Since $2 \leq x < 3$, we begin by letting

$x_0 = e - 2 = 0.71828 \ldots$ as in our proof of Proposition 3. In this case, $n = 7$, and so we locate where x_0 lies in the Farey sequence F_7. We find readily that $5/7 = 0.71428 \ldots < x_0 < \frac{3}{4} = 0.75$. The mediant of $5/7$ and $3/4$ is $8/11 = 0.7272 \ldots$ Since x_0 lies between $5/7$ and $8/11$, we let $p_0 = 5$ and $q_0 = 7$. Finally, let $mq_0 + p_0 = 2(7) + 5 = 19$ and $q = q_0 = 7$. So our answer is $19/7$. As a check, note that

$$|e - 19/7| < 0.004001 < \frac{1}{7(8)} = \frac{1}{56} = 0.17857\ldots.$$

Next, let's look at a couple of cool properties exhibited by Farey fractions. First, if we add up all the fractions in any sequence of Farey fractions, the sum will be half of the number of fractions in that row. For example, consider F_3: $0/1, 1/3, 1/2, 2/3, 1/1$. There are five fractions in the sequence F_3 and

$$\frac{0}{1} + \frac{1}{3} + \frac{1}{2} + \frac{2}{3} + \frac{1}{1} = 5/2.$$

Check this on the sequences F_4 and F_5.

The second property is this: The sum of all the denominators in any Farey sequence is twice the sum of all the numerators. Again, look at F_3. In this case, the sum of the numerators is $0 + 1 + 1 + 2 + 1 = 5$. The sum of the denominators is $1 + 3 + 2 + 3 + 1 = 10$, which is twice the sum of numerators. Again, check this out on the sequences F_4 and F_5.

Why is this true? Begin with the second property. If $\frac{a}{b}$ is an element of F_n, then

$$1 - \frac{a}{b} = \frac{b - a}{b}$$

is an element of F_n as well. We can pair up these two fractions. The sum of their numerators is $a + (b - a) = b$, while the sum of their denominators is $b + b = 2b$. Care must be taken by noting that the fraction $\frac{1}{2}$ has no mate in this scheme. So we add up all the rest and finally add 1 to the numerator and 2 to the denominator, still maintaining the same ratio between numerator and denominator.

Similarly, the first property follows by doing the same matching process for all the fractions except ½. For each pair of fractions, their sum is

$$\frac{a}{b} + \left(1 - \frac{a}{b}\right) = 1.$$

It takes two fractions to add to 1. Finally, add in the last fraction ½. It is still the case that the sum of all the Farey fractions will be half of the number of such fractions.

Finally, I'd like to explain a little discovery I made myself about Farey fractions. To ensure that our Farey sequences were unambiguously well defined, our Farey sequences contained reduced fractions only. The fraction 1/3 was written only as 1/3 rather than allowing any equivalent representation such as 2/6 or 3/9. Note that the definition of mediant does not actually require that the fractions a/b and c/d necessarily be reduced fractions. It is still the case that the mediant of any two fractions will be a rational number lying strictly between them. However, even if

$$\frac{A}{B} = \frac{a}{b}.$$

with $A \neq a$, the mediant

$$\frac{A}{B} \oplus \frac{c}{d}$$

will not equal the mediant

$$\frac{a}{b} \oplus \frac{c}{d}.$$

For example, the mediant of 1/3 and 1/2 is 2/5 while the mediant of 1/3 and 2/4 is 3/7 even though ½ = 2/4. By rewriting fractions so that their numerators and denominators are not necessarily relatively prime (i.e., so that the fraction may no longer be in reduced form), it seems natural to wonder what intermediate fractions can be created as their mediants. The answer turns out to be *all of them*! In fact, for any two distinct rational numbers, all intermediate fractions are expressible as the mediant of some representation of the two given rationals. Let's state this as a proposition and then give a constructive proof. We'll make use of what we learned about matrices in Chapter 7. Afterwards, we'll check out some examples.

Proposition 4: *Let*

$$\frac{a}{b} < \frac{c}{d}.$$

If

$$\frac{a}{b} < \frac{f}{g} < \frac{c}{d},$$

then there are equivalent fractions

$$\frac{A}{B} = \frac{a}{b} \text{ and } \frac{C}{D} = \frac{c}{d}.$$

such that

$$\frac{f}{g} = \frac{A+C}{B+D} = \frac{A}{B} \oplus \frac{C}{D}.$$

Proof of Proposition 4: It suffices to find rational numbers j and k having a common denominator such that

$$f = ja + kc \text{ and } g = jb + kd.$$

In this case, we than have that $A = ja$, $B = jb$, $C = kc$, and $D = kd$. This linear system of equations can be written as a matrix equation $AX = B$ where

$$A = \begin{bmatrix} a & b \\ c & d \end{bmatrix} \text{ and } X = \begin{bmatrix} j \\ k \end{bmatrix}, \text{ and } B = \begin{bmatrix} f \\ g \end{bmatrix}.$$

Since $\frac{a}{b} \neq \frac{c}{d}$, $ad - bc \neq 0$ and hence A is invertible. Thus, X = $A^{-1}B$. Explicitly, utilizing our previous result on inverses of 2×2 matrices, $x = \begin{bmatrix} j \\ k \end{bmatrix}$, where

$$j = \frac{df - cg}{ad - bc}$$

and

$$k = \frac{ag - bf}{ad - bc}.$$

Although $f = A + C$ and $g = B + D$, the numbers j and k need not be integers. Hence, clear fractions by multiplying both by $ad - bc$. We then redefine j and k by $j = df - cg$ and $k = ag - bf$.

Example: Let $\frac{a}{b} = \frac{1}{2}$ and $\frac{c}{d} = \frac{2}{3}$. Find fractions equal to 1/2 and 2/3 whose mediant is $\frac{53}{100}$.

Solution: In this case, $j = df - cg = -41$ and $k = ag - bf = -6$. Hence,

$$\frac{1}{2} = \frac{A}{B} = \frac{41}{82} \text{ and } \frac{2}{3} = \frac{C}{D} = \frac{12}{18}.$$

(We simplified by multiplying j and k by -1.) Note that $df - cg < 0$ and $ag - bf < 0$. In practice, to obtain positive factors j and k, simply rewrite the inequality as

$$\frac{c}{d} > \frac{f}{g} > \frac{a}{b}$$

and then let

$$j = \det \begin{bmatrix} c & f \\ d & g \end{bmatrix}$$

and

$$k = \det \begin{bmatrix} f & a \\ g & b \end{bmatrix}.$$

Example: Let

$$\frac{a}{b} = \frac{2}{3}, \frac{f}{g} = \frac{2}{1}, \text{ and } \frac{c}{d} = \frac{9}{2}.$$

Then

$$j = \det \begin{bmatrix} 9 & 2 \\ 2 & 1 \end{bmatrix} = 5 \text{ and } k = \det \begin{bmatrix} 2 & 2 \\ 1 & 3 \end{bmatrix} = 4.$$

Thus,

$$\frac{2}{3} = \frac{10}{15}, \frac{9}{2} = \frac{36}{8},$$

and the mediant of $\frac{10}{15}$ and $\frac{36}{8}$ is

$$\frac{46}{23} = \frac{2}{1}.$$

The mediant of two numbers is usually not the mean average of them. However, by Proposition 4, we can always find fractions equivalent to a/b and c/d having mediant equal to the average of them. Simply let $j = d$ and $k = b$.

In our first example with $\frac{a}{b} = \frac{1}{2}$ and $\frac{c}{d} = \frac{2}{3}$, let $j = 3$ and $k = 2$. Then $\frac{1}{2} = \frac{3}{6}$ and $\frac{2}{3} = \frac{4}{6}$. The mediant

$$\frac{3}{6} \oplus \frac{4}{6} = \frac{7}{12},$$

which is the mean average of the original fractions.

Exercises

1. (a) Determine F_7, the Farey fractions of order 7.
   (b) Determine F_8, the Farey fractions of order 8.
2. Verify Propositions 1 and 2 on some examples in F_7 and F_8.
3. Find a fraction p/q such that

$$\left| \sqrt{2} - \frac{p}{q} \right| < \frac{1}{8q},$$

where q satisfies $1 \le q \le 7$.

4. Find a fraction p/q such that

$$\left| \sqrt{5} - \frac{p}{q} \right| < \frac{1}{8q},$$

where q satisfies $1 \le q \le 7$.

5. Find a fraction p/q such that

$$\left| \sqrt{5} - \frac{p}{q} \right| < \frac{1}{10q},$$

where q satisfies $1 \le q \le 9$.

6. Find a fraction p/q such that

$$\left| e^2 - \frac{p}{q} \right| < \frac{1}{8q},$$

where q satisfies $1 \le q \le 7$.

7. (a) Find fractions $a/b = 1/3$ and $c/d = 1/2$ such that $a/b \oplus c/d = 9/20$.

   (b) Find fractions $a/b = 1/3$ and $c/d = 1/2$ such that $a/b \oplus c/d = 3/8$.

8. (a) Find fractions $a/b = 1/4$ and $c/d = 1/3$ such that $a/b \oplus c/d = 3/10$.

   (b) Find fractions $a/b = 1/4$ and $c/d = 1/3$ such that $a/b \oplus c/d = 7/24$.

Chapter 9

A Digression on Cutting Cakes and Marking Sticks

> The advantage of growing up with siblings is that you get very good at fractions.
>
> **Robert Brault, Round Up the Usual Suspects (2014)**

So far, our fractions have been abstract entities. In this interlude, let's talk about fractions that really matter—namely, getting your fair share of a cake. Suppose two people, say a boy and a girl, want to share a cake. For example, think of a rich, round, yummy chocolate birthday cake! How can they slice it so that each of them is confident they've gotten a proper share? The solution is simple. Let one of them, say the girl, cut the cake so that she's sure both pieces are exactly half the cake. Then let the boy choose either piece. If he thinks one is slightly larger, then he'll choose that one. Otherwise, he just picks either piece. Finally, the girl gets the remaining piece and both are satisfied with their share.

But what if there are more than just two people? Now the situation is quite a bit more complicated. Even with just three people, you can envision the first person cutting off more than half the cake for a piece. Then whoever gets that piece will leave the others feeling cheated. This problem of fairly apportioning the slices has many avenues to investigate. There are a multitude of interesting solutions to the problem with various assumptions and conditions on number of cuts, the sizes and shapes of cuts, whether the portions have to be just one piece, whether the pieces need be connected, etc. Here we describe a method that works well for three people followed by two very simple procedures that will work with any number of people.

Suppose Alice, Bob, and Clair wish to share a cake equitably. Alice cuts the cake into three equal pieces (by her reckoning at least) which we denote pieces p, q, and r. Bob selects what he perceives as the two biggest pieces (say p and q). Let's say he perceives that p is the largest and q second largest. He carefully evens them out if necessary, by taking an ε amount from p and adding it to q. We now have three pieces of cake, namely $p - \varepsilon$, $q + \varepsilon$, and r. Next, Clair gets to choose any piece she wishes from among them satisfied that she's receiving at least a third of the cake. If she chooses piece r, then Alice gets $q + \varepsilon$ and Bob gets $p - \varepsilon$. Everyone is happy. If she chooses piece $p - \varepsilon$, then Bob gets $q + \varepsilon$, Alice is left with piece r, and everyone is content once again. Finally, if Clair chooses piece $q + \varepsilon$, then Bob gets $p - \varepsilon$, Alice gets r, and all is well.

In the exercises, we'll extend our reasoning to the case with four people. But as the number of people increases, the solution seems to require ever greater complexity. Fortunately, there are some easily described solutions that work well with an arbitrary number of people. We consider two general methods.

Method 1: Number the people $1, 2, \ldots, n$ ordered in any way you wish (age, height, alphabetical by last name, etc.) We'll use female pronouns throughout this example. Person 1 cuts one slice out of the cake that she considers to be precisely $1/n$ of the entire cake— a portion she considers to be entirely fair. Person 2 then can strip away (diminish) a bit of that slice which still leaves a fair slice ($1/n$ of the entire cake). Person 2 can also choose not to touch the slice if she considers the slice to be exactly $1/n$ of the entire cake or smaller. Now Person 3 gets the same choice with that original slice—cut off a bit more or leave it alone. We continue this process for all n people. At that point, the last person to cut into the slice gets to keep it. It may well be that person 1 gets to keep her original slice or it may be someone else. But in any event, whoever gets the slice will be happy that they've received their fair share. Everyone else will consider the remainder of the cake to contain more than $1 - 1/n = (n - 1)/n$ of the original cake. There are $n - 1$ people remaining yet to get some cake. Let the first person still remaining begin the same process that we discussed above. The rest of the people follow suit and a second piece will then go to the last person to diminish the second slice. Now repeat until all *n* people have a slice. Everyone will be content that they've received at least their fair share of the cake. Try it and see!

Method 2: Again, assume we've numbered the participants $1, 2, \ldots, n$. We'll use male pronouns this time. Person 1 starts by taking the cake and dividing it exactly in half as best he can. He's now confident that either piece yields a full half of the cake. Next Person 2 chooses one of the pieces. If he thinks they're both exactly half, there's no problem. If he thinks one piece is slightly larger, then he'll choose the larger piece. Either way, Persons 1 and 2 are content with the current state of affairs. If there is a third person, then Persons 1 and 2 cut each of their pieces into three equally-sized slices. Person 3 chooses what he considers the largest of the three slices from each of the shares belonging to the first two people. Persons 1 and 2 are still satisfied that they have at least a third of the cake while Person 3 has at least one-third of each of their shares. Together those two pieces will add up to a third or more of the entire cake. We can continue this process. In general, for the *n*th person, Persons $1, 2, \ldots, n - 1$ all cut their cake into *n* equal pieces as best they can determine, then Person *n* picks the largest of the slices from each of the first $n - 1$ people. Everyone should be satisfied that they have at least their fair share. Of course, they might not be so excited about having what may amount to a dish full of cake crumbs, but that's what they get for being so fussy in the first place!

Now let's move on to a sticky problem. Well, actually a problem dealing with a straight stick. We make two marks on it randomly and independently. If we then take the stick and break it into *k* equal-lengthed pieces for some $k > 1$, what is the probability that the marks lie on the same piece?

The probability of an event is its likelihood of happening and is always some number between 0 and 1 inclusive. Of course, most of these numbers aren't rational numbers, but the answer to the question above certainly is. The answer is $1/k$. The first mark is made randomly and the second mark has an equal chance of being on any of the *k* equal-lengthened pieces.

For a more interesting question, suppose after making the two marks that we break the stick randomly into k pieces of possibly variable length. What is the probability now that the two original marks end up on the same piece? Is the answer still $1/k$, less than $1/k$, more than $1/k$? You may want to ponder this a bit before reading on.

The general case involves $k+1$ random points (the two marks plus the $k-1$ break points where the stick was broken). Hence there are

$$\binom{k+1}{2} = \frac{k(k+1)}{2}$$

choices for the position of the two marks. If we list the random points in order from bottom of the stick to top, the two marks end up on the same piece if and only if they are consecutive points. There are k such consecutive pairs in all. Hence the answer is

$$\frac{k}{\binom{k+1}{2}} = \frac{2}{k+1}.$$

So the chance of the markings being on the same piece in this randomly divided stick is greater than in the uniformly broken case—in fact, nearly twice as likely! Instead of dividing a stick, if we are dealing with voting districts or economic or resource issues, the stakes might be somewhat higher.

If you're still "sticking around," here is a related problem discussed by the master mathematical expositor Martin Gardner in his collection *Mathematical Puzzles and Diversions*. Once again choose two points randomly and independently on a stick and then break the stick at those points to make three smaller pieces. What is the probability that the three pieces can be arranged to form a triangle? Of course, this is equivalent to asking for the probability that no piece is longer than the sum of the lengths of the other two pieces.

By way of background, we require an important geometric result. Viviani's theorem states that for any point P inside an equilateral

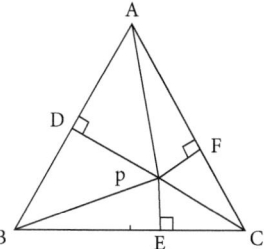

Figure 9.1 Diagram illustrating Viviani's theorem.

triangle, the sum of the perpendiculars to the three sides equals the height of the triangle (see Figure 9.1). Vincenzo Viviani (1622–1703) translated and restored several important geometric works of the ancient Greeks. He also lived with and assisted the blind Galileo during his last few years. Afterwards, he wrote the only extant biography of Galileo due to a contemporary.

In Figure 9.1, let $\triangle ABC$ be an equilateral triangle with side length a. Choose any point P inside it and draw perpendicular line segments to each side of lengths PD, PE, and PF. $\triangle ABC$ is partitioned into three triangles $\triangle APB$, $\triangle APC$, and $\triangle CPB$. The sum of the areas of the three smaller triangles equals the area of $\triangle ABC$. Since the area of a triangle is one-half its base times height,

$$\frac{a \cdot PD}{2} + \frac{a \cdot PE}{2} + \frac{a \cdot PF}{2} = \frac{ha}{2}.$$

Hence, $PD + PE + PF = h$, independent of our choice of P.

To solve the problem of whether our three sticks can form a triangle, we consider the following diagram, Figure 9.2, where $\triangle ABC$ is an equilateral triangle once again and the points G, H, and I bisect its three sides.

In Figure 9.2, the triangles $\triangle AGI$, $\triangle GBH$, $\triangle GIH$, and $\triangle HIC$ are all congruent equilateral triangles and thus each are of area one-fourth that of $\triangle ABC$. Pick any point P within $\triangle ABC$ as in Figure 9.1. We have shown that the sum $PD + PE + PF$ is the same no matter which point P is chosen.

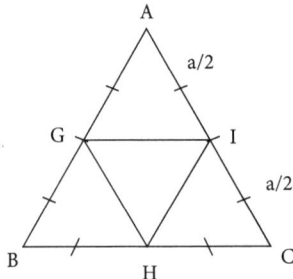

Figure 9.2 Four congruent equilateral triangles within $\triangle ABC$.

Now let our original stick be of length h, the height of $\triangle ABC$. There is a one-to-one correspondence between the points P within $\triangle ABC$ and lengths PD, PE, and PF representing the lengths of the three random pieces. The pieces can be arranged to form a triangle if and only if P lies in $\triangle GIH$ since other points necessarily have one length greater than $h/2$ and thus are longer than the sum of the lengths of the other two pieces. So the probability that a triangle can be formed from the three pieces is precisely 1/4. A variant of this problem is discussed in this chapter's Exercises.

Now let's consider an odd pizza problem taken from the first issue of the journal *Quantum* (January 1990) which was the English language version of the undergraduate Soviet (later Russian) physics and math journal *Kvant*. We now consider a pizza in the shape of an equilateral triangle. Okay, I said this was going to be a bit odd. Pick any point P inside it and make six slices by cutting from the chosen point to each of the three vertices and from the chosen point orthogonally (perpendicularly) to each side. Show that if two people consume the pizza by eating alternate slices, then each gets exactly half the pie.

Solution: The key is to add in three additional line segments through P parallel to the three sides of the triangle (see Figure 9.3).

Moving counterclockwise from the bottom left of the pizza, the slices can be labeled A, B, C, D, E, and F with the first person receiving slices A, C, and E and the second person receiving slices B,

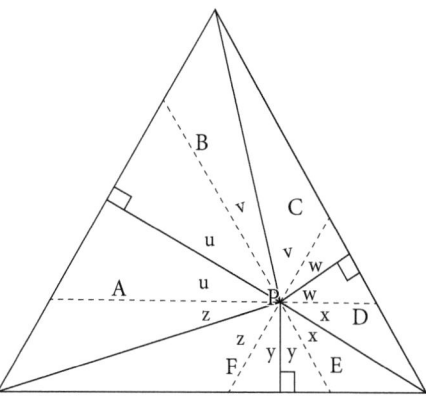

Figure 9.3 Sum of alternate slices have equal area.

D, and F. The slices and line segments divide slice A into parts z and u, slice B into u and v, slice C into v and w, slice D into w and x, slice E into x and y, and slice F into y and z. This follows from the fact that the v's, x's, and z's bisect parallelograms while the u's, w's, and y's bisect isosceles triangles. Hence, the sum of the areas

$$
\begin{aligned}
A + C + E &= (z + u) + (v + w) + (x + y) \\
&= (u + v) + (w + x) + (y + z) \\
&= B + D + F.
\end{aligned}
$$

Getting hungry? Here's a riddle: What's the difference between a mathematician and a large pizza? Answer: A large pizza can feed a family of four.

Finally, here is a classic brain teaser involving fractions of a brick of gold. It can be solved without any technical mathematics. Suppose you have a brick of gold of uniform thickness (say of dimensions 1 unit by 2 units). If the entire brick is worth seven days' work, where should you make two cuts through it so that you could pay a worker for any number of days from 1 to 7?

Once you realize that you need to be able to have a piece worth one-seventh of the entire brick and that you also need to be able to pay a worker for exactly two days' work as well, there aren't many possibilities to ponder. Below is the solution where the dimensions are given in our chosen units. Please note, for example, that a slice of area $1 \times \dfrac{2}{7} = \dfrac{2}{7}$ constitutes one-seventh of the area of the original brick (see Figure 9.4).

Of course, the solution can be easily generalized so that with three cuts an appropriate golden brick can be subdivided into four pieces so that any integral number of days from 1 to 15 could be paid (see Figure 9.5).

Interestingly, there is actually a way to make just two cuts rather than three to subdivide the golden brick into pieces so that any integral number of days from 1 to 15 could be paid. You don't have to think "outside the box," but you do have to consider making cuts that are orthogonal to each other rather than parallel. Below is the solution (see Figure 9.6).

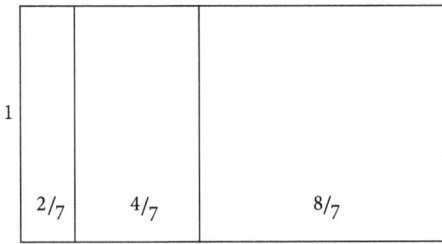

Figure 9.4 Brick sliced to pay any number of days from 1 to 7.

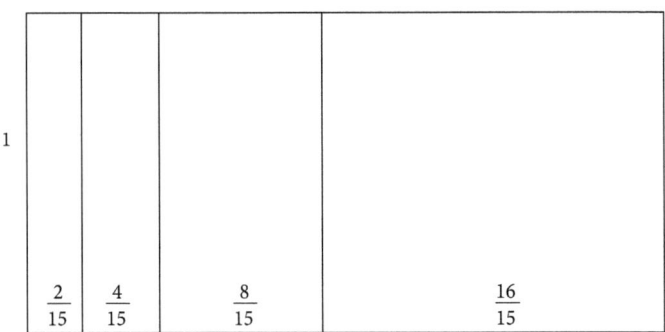

Figure 9.5 Brick sliced to pay any number of days from 1 to 15.

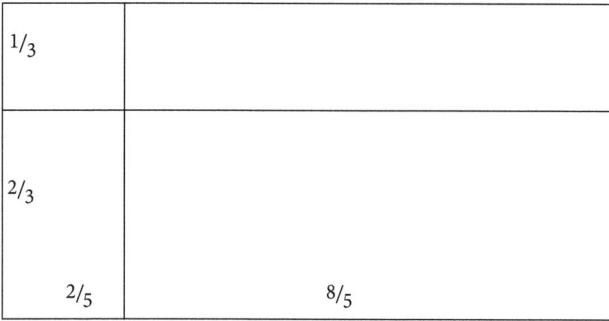

Figure 9.6 Two slices can pay any number of days from 1 to 15.

Exercises

1. What is the largest number of pieces a large circular pizza be cut into with n straight cuts (chords of the circle)? How many include a piece of the crust?

2. (a) Wages for a part day's job can be any dollar amount from $1 to $127. What amounts can be placed in seven separate wallets so that any whole dollar amount can be distributed by choosing appropriate wallets?

   (b) What if there are ten wallets covering any whole dollar amount from $1 to $1023?

3. If a is the length of side of an equilateral triangle, show that its height is

$$h = \frac{a\sqrt{3}}{2}.$$

4. If a stick is broken into three pieces by first snapping it in two, then randomly, arbitrarily choosing one of the pieces and randomly snapping that in two, show that the probability that the three remaining pieces can form a triangle is now just 1/6. Is it surprising that this differs from our earlier result?!

5. Here is a solution for dividing a cake among four friends—Alice, Bob, Clair, and Dan. (Please fill in any details needed.)

Alice, Bob, and Clair follow the procedure that we discussed previously for fair division among three people. At the conclusion of their procedure, each is confident that they possess at least a third of the cake. Next, Alice, Bob, and Clair cut their portions into four equal pieces so that each considers their pieces to be at least one-twelfth of the entire cake. Dan next chooses the biggest slice from each of his three friends. Alice, Bob, and Clair then retain their remaining pieces. Everyone is happy!

6. How can a large rectangular brick of gold be subdivided with four parallel cuts so that any number of days' wages during the month of July can be paid?

7. What is the relationship between the solutions to the gold brick problems and binary arithmetic?

Chapter 10
Egyptian Fractions

All is number.

Pythagorean motto (6th century BCE)

The ancient Egyptians of Africa have left us with a rich historical record of their civilization which thrived along the banks of the Nile River for several millennia. First there was a Nubian society (now in Sudan), followed by the Old Kingdom (3500–2000 BCE), Middle Kingdom (2000–1600 BCE), and New Kingdom periods (1600–1100 BCE). What has come down to us over all these millennia are many enormous monuments (including the incredible pyramids of Giza and elsewhere) with their hieroglyphics, "sacred carvings," and over a thousand papyri with either hieratic "sacred writing" pen and ink cursive writing or, much later, a demotic or popular writing script. Much of the writing involves tributes to their pharaohs and fanciful historic accounts of their various gods and goddesses (of which they had nearly 2000—some of which appear in many forms). None of this could be deciphered until the discovery of the Rosetta Stone in 1799 during the Napoleanic "Expedition" to Egypt. The stone, a granodiorite stele measuring $3'8'' \times 2'6'' \times 11''$, had a long message written in three separate panels—the top a hieroglyphic (language of the gods), the middle a demotic or cursive Egyptian script, and the bottom portion in ancient Greek. Even so, it took twenty-three years to fully decipher the text—accomplished by a team led by Jean Francois Champollion (1790–1832), professor from Grenoble who was well versed in Greek, Coptic, and other ancient languages.

In addition to the vast monuments with their hieroglyphs, there is an abundance of hieratic records (pen and ink) preserved on papyrus strips. Among these are several dating to approximately

1850–1650 BCE dealing exclusively with mathematical matters for practical use. Specific numerical problems were posed with detailed solutions concerning feed mixtures for cattle and domestic fowl, strength of beer, storage of grain, taxes, and food allotments.

Pharaonic Egypt is the first civilization that we know of that used a 365-day annual calendar consisting of twelve 30-day months and a 5-day holiday period at the end of the year. (Seem familiar?) Like all astronomically sophisticated societies, they were well aware that they could only approximate the ratio between the length of a day (single rotation of the Earth), a month (a revolution of the moon about the Earth), and a year (Earth's single revolution about the sun). Of course, their conception of the causes of a day and year were more geocentric. A day was divided into 12 hours and each night 12 hours with the hours being of variable duration depending on the time of year.

For any society to create a reliable calendar, whether lunar, solar, or lunar/solar, it is often essential that they grapple with fractions and with approximating irrational quantities with them! Our current Gregorian calendar is just one more imperfect example of this.

The ancient Egyptians noticed that the flooding of the Nile took place about every 365 days when the star Sirius rose in the east just before sunrise. Since the year is actually much closer to 365 and a quarter days (though still just approximately), their calendar was only correct once about every 1460 years = 365 × 4. Since the Roman scholar Censorinus noted that the calendar was in alignment with the seasons in the year 139 CE, by extrapolating backwards either two or three cycles, the calendar may have been instituted around 2780 or 4240 BCE give or take. Nice observation, Censorinus!

Most of our current knowledge of ancient Egyptian mathematics comes from two main sources written in hieratic. The Moscow mathematical papyrus (appx. 1850 BCE) consists of twenty-five mathematical problems written on a papyrus 18 feet long and 3 inches wide (with some variation). This was purchased by a Russian antiquities dealer in Thebes in 1892 and is now in the Pushkin Institute Museum in Moscow. The other papyrus is the Rhind mathematical papyrus (appx. 1650 BCE) written by a scribe named

Ahmes, who copied it from an earlier text. It is 18 feet long and 1 foot wide. Its title translates as "Accurate reckoning for inquiring into things, and the knowledge of things, mysteries, and all dark secrets." The papyrus itself was illegally taken by the Scottish antiquary Henry Rhind near Luxor in 1858 and then was sold to the British Museum in 1865 (where it resides to this day).

The Egyptians used a base 10 numeral system with a different pictograph for each successive power of 10 from 1 up to 10,000,000. For example, 1 was represented by a vertical stroke, 10 by a heel bone, a snare (or coil of rope) for 100, a lotus flower for 1,000, a bent finger for 10,000, a burbot fish for 100,000, the god Heh with upraised arms for 1,000,000, and so on (Figure 10.1). Repetition was used. Two lotus flowers followed by five vertical strokes represented 2005.

To the ancient Egyptians, the proper concept of a fraction was a fraction with 1 as the numerator. In hieroglyphs, this was denoted by placing the symbol for a mouth directly over the denominator. In cursive hieratic writing, the mouth was replaced with an oval (and even later, simply with a dot). Other fractions were handled via such sums of distinct unit fractions. We call all such sums *Egyptian fractions*. For example, the fraction $\frac{5}{7}$ might be written as $\frac{1}{2} + \frac{1}{5} + \frac{1}{70}$. To us, this might seem unnecessarily cumbersome. But it had some practical applications. For example, suppose you wanted to split five

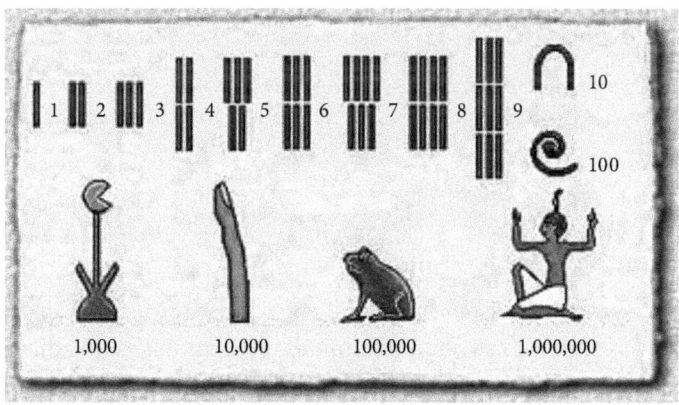

Figure 10.1 Egyptian numeral hieroglyphs.
Reproduced from BbcNkl (2015) via Wikimedia Commons.

equally sized loaves of bread among seven individuals. One could divide each loaf into seven equal pieces and then give everyone five pieces of the sliced-up loaves, each piece one-seventh of a loaf. This would be completely fair. But perhaps some individuals wished to have a larger chunk of the loaf included in their share. In that case, four of the loaves could be cut in half instead. Seven of the halves could be evenly distributed. The last loaf could then be cut into fifths. In addition, two fifths could be cut from the last half loaf remaining. Everyone would then get an additional fifth of a loaf. The remaining tenth of a loaf could be evenly distributed or simply tossed or maybe just fed to the birds. In this way, everyone would have a more substantial and perhaps more usable chunk of bread.

The Rhind papyrus contains a table for writing fractions of the form $2/n$ for odd n from 5 to 101. For example, $\frac{2}{5} = \frac{1}{3} + \frac{1}{15}$ and $\frac{2}{7} = \frac{1}{4} + \frac{1}{28}$. For a given fraction, there are many possible ways to rewrite it as a sum of unit fractions. Note, for example, that if we allow $\frac{1}{n}$ to be one of the terms, then

$$\frac{2}{n} = \frac{1}{n} + \frac{1}{2n} + \frac{1}{3n} + \frac{1}{6n}.$$

Yet that identity is only used sporadically within the Egyptian papyri that have come down to us. More often, they utilized the identity

$$\frac{2}{n} = \frac{1}{\frac{n+1}{2}} + \frac{1}{\frac{n(n+1)}{2}}.$$

And there is nothing special about just fractions of the form $2/n$. For example,

$$\frac{3}{7} = \frac{1}{3} + \frac{1}{14} + \frac{1}{42} \text{ as well as } \frac{3}{7} = \frac{1}{4} + \frac{1}{6} + \frac{1}{84}.$$

To check your understanding, see if you can write the fractions $\frac{2}{9}$, $\frac{3}{11}$, and $\frac{109}{120}$ as sums of Egyptian fractions with all terms having different denominators.

Here is problem 63 from the Rhind papyrus. Divide 700 loaves of bread among four people in the continued proportion 2/3: 1/2: 1/3: 1/4.

Solution: The problem itself seems nonsensical. The proportions allotted are too large. They should add up to 1. Apparently, the idea is that the four people get 2/3, 1/2, 1/3, and 1/4 of something which totals 700 loaves. But

$$2/3 + 1/2 + 1/3 + 1/4 = 7/4.$$

So take 4/7 of 700 which is 400 as our "unit." Then we get (2/3)400 = 266 2/3, (1/2)400 = 200, (1/3)400 = 133 1/3, and (1/4)400 = 100 loaves as the four shares.

Here is a more challenging problem from the Rhind Papyrus: Divide 100 loaves among five people in such a way that the shares received are in arithmetic progression and 1/7 of the sum of the largest three shares equals the sum of the smallest two shares!

Even in ancient times, there was some interest in mathematics for its own sake rather than just as a practical tool.

Solution: In order that the shares be in arithmetic progression, let the shares be

$$a, \ a+b, \ a+2b, \ a+3b, \ a+4b,$$

where

$$a + (a + b) + (a + 2b) + (a + 3b) + (a + 4b) = 100$$

and

$$(1/7)\,[(a + 2b) + (a + 3b) + (a + 4b)] = a + (a + b).$$

The first equation leads to $5a + 10b = 100$ or $a + 2b = 20$. Thus,

$$a = 20 - 2b. \tag{10.1}$$

Multiplying both sides by 7 in the second equation yields

$$3a + 9b = 14a + 7b.$$

This is equivalent to

$$11a = 2b \qquad\qquad (10.2)$$

once we combine the a's and the b's.

Substituting the value for a in equation (10.1) into equation (10.2) gives us $11(20-2b) = 2b$ which implies that $24b = 220$. Simplifying,

$$b = 55/6 \text{ and } a = 5/3.$$

Therefore, the final answer is that shares be distributed as

$$5/3 + (55/6)n \text{ for } n = 0, 1, 2, 3, 4.$$

So the appropriate shares as mixed fractions are

$$1\frac{2}{3}, 10\frac{5}{6}, 20, 29\frac{1}{6}, \text{ and } 38\frac{1}{3} \text{ loaves}.$$

You might wonder if there is always a consistent way to rewrite a common fraction by a sum of unit fractions. One natural tack you could take is known as the *greedy algorithm*. At each step of the way, remove the largest unit fraction possible. For example, the largest unit fraction less than 3/7 is 1/3 since the next unit fraction 1/2 is larger than 3/7. Hence, we write $\frac{3}{7} = \frac{1}{3} + \frac{2}{21}$.

Next, 1/11 is the largest unit fraction less than 2/21. But

$$\frac{2}{21} = \frac{1}{11} + \frac{1}{231}.$$

Thus,

$$\frac{3}{7} = \frac{1}{3} + \frac{1}{11} + \frac{1}{231}.$$

Compare with two other representations for 3/7 that we mentioned previously.

For the fraction $\frac{53}{72}$, the greedy algorithm results in the sum

$$\frac{53}{72} = \frac{1}{2} + \frac{1}{5} + \frac{1}{28} + \frac{1}{2520}.$$

It might not even be clear that this algorithm always works. In other words, could the sum go on forever always leaving over some positive quantity that's not itself a unit fraction? And does the sum always give us the fewest terms or those with the smallest denominators? You can double-check that in addition to our sum above; alternatively, we can write

$$\frac{53}{72} = \frac{1}{2} + \frac{1}{8} + \frac{1}{9}.$$

Thus, the greedy algorithm doesn't always lead to a solution with either the fewest terms or the smallest denominators.

However, the greedy algorithm will always terminate at least. Thus, we can rest assured that it always produces a valid answer. We state this as a proposition and give a quick proof.

Proposition 1: *Given a fraction $\frac{a}{b}$ with a < b, the greedy algorithm will terminate in at most a steps.*

Proof: If $\frac{a}{b}$ is equivalent to a unit fraction, then the result holds. Assume that $\frac{a}{b}$ is not a unit fraction. The first step is to find the value of n for which

$$\frac{1}{n+1} < \frac{a}{b} < \frac{1}{n}.$$

In this case, the fraction $\frac{1}{n+1}$ will be the first term of our sum.

Notice that $b > an$ for the second inequality to hold.

When we subtract $\frac{1}{n+1}$ from $\frac{a}{b}$, the remainder is

$$\frac{a}{b} - \frac{1}{n+1} = \frac{a(n+1) - b}{b(n+1)} > 0.$$

Our new numerator is $a(n+1) - b$. But $b > an$ implies that $an - b < 0$ and hence $a(n+1) - b = a + (an - b) < a$. Thus, the new numerator is smaller than the original one. Similarly, at each step of our algorithm, the remaining fraction will have ever decreasing numerators. But there are less than the number a of positive integers less than a. So eventually, our algorithm will terminate after a finite number of steps resulting in a sum of unit fractions as desired.

After all these years, finding the shortest sum is a problem still not fully understood. In 1948, Paul Erdös and Ernst Straus conjectured that the equation

$$\frac{4}{n} = \frac{1}{x} + \frac{1}{y} + \frac{1}{z}.$$

is always solvable for $n > 1$ (where x, y, z need not be distinct). Our proposition guarantees that $\frac{4}{n}$ can be expressed as a sum of at most four distinct unit fractions, but they conjectured that only three unit fractions are required (with possible repetition). In the Exercises, we'll prove a special case of this conjecture. However, a full proof or refutation of their conjecture remains unresolved although it's been verified for all $n > 1$ up to $n = 10^{14}$. In addition, the Ethiopian mathematician Dagnachew Jenber Negash (2019) has proven its veracity for all n except a sparse subset of $n \equiv 1 \pmod 8$.

Let's investigate ways to write the number 1 as a sum of Egyptian fractions. Of course,

$$1 = \frac{1}{2} + \frac{1}{2} = \frac{1}{3} + \frac{1}{3} + \frac{1}{3} = \ldots = \frac{1}{n} + \ldots + \frac{1}{n} = \ldots$$

for all $n \geq 2$ where the nth sum has n terms. But what if we insist on sums of *unique* distinct unit fractions?

We begin with the observation that 1 can be expressed as the sum of three such fractions

$$1 = \frac{1}{2} + \frac{1}{3} + \frac{1}{6}.$$

Next, verify the identity

$$\frac{1}{n} = \frac{1}{n+1} + \frac{1}{n(n+1)}.$$

Utilizing this identity, we can replace the last term (with the largest denominator) to obtain an equation with one more term. In this way, we have shown that 1 can be expressed as a sum of n distinct unit fractions for any $n \geq 3$. For example,

$$1 = \frac{1}{2} + \frac{1}{3} + \frac{1}{7} + \frac{1}{42} = \frac{1}{2} + \frac{1}{3} + \frac{1}{7} + \frac{1}{43} + \frac{1}{1806}, \text{ etc.,}$$

where $1806 = 42 \times 43$.

As an interesting variant, the question of how close to 1 a sum of unit fractions can be without quite attaining 1 is nicely discussed in the book *Which Way Did the Bicycle Go?*, by Joseph Konhauser, Dan Velleman, and Stan Wagon. Of course, ½ is the closest single unit fraction less than 1. What about the sum of two unit fractions? Again, it's easy to see that

$$\frac{1}{2} + \frac{1}{3} = \frac{5}{6}.$$

comes closer to 1 than any other such addition. Next, consider the inequality

$$\frac{1}{a} + \frac{1}{b} + \frac{1}{c} < 1. \tag{10.3}$$

We seek distinct positive integers a, b, and c for which the above sum is as close to 1 as possible without reaching 1.

To cut to the chase, the answer is

$$\frac{1}{2} + \frac{1}{3} + \frac{1}{7} = \frac{41}{42}.$$

Let's see why.

Without loss of generality, let's assume $a < b < c$, which is consistent with the ascending order of denominators we've used all along. If $a \geq 3$, then the largest sum possible is 47/60. But this comes up 13/60 short, whereas

$$\frac{1}{2} + \frac{1}{3} + \frac{1}{7}.$$

is just 1/42 less than 1. Hence, $a = 2$. Next, if $b \geq 4$, then the best we can do is

$$\frac{1}{2} + \frac{1}{4} + \frac{1}{5} = \frac{19}{20}.$$

which is still 1/20 less than 1. Hence $a = 2$ and $b = 3$. Then it must be the case that $c \geq 7$ with optimal value being $c = 7$.

The situation gets more complicated as we tack on additional terms. Even so, consider

$$\frac{1}{a} + \frac{1}{b} + \frac{1}{c} + \frac{1}{d} < 1. \tag{10.4}$$

As before, we seek distinct positive integers a, b, c, d for which the above sum comes as close to 1 as possible. The analysis is more intricate, but follows the same sort of pattern as that for inequality (10.3).

The closest a sum of four unit fractions can get to 1 without reaching or surpassing it is

$$\frac{1}{2} + \frac{1}{3} + \frac{1}{7} + \frac{1}{43} = \frac{1805}{1806}. \tag{10.5}$$

Here is the verification: If $a \geq 3$, then the largest the sum could be is

$$\frac{1}{3} + \frac{1}{4} + \frac{1}{5} + \frac{1}{6} = \frac{19}{20}.$$

This is close to 1, but not nearly as close as $\frac{1805}{1806}$. Hence, $a = 2$.

Next, $b < 6$, since if $b \geq 6$, the largest sum would be

$$\frac{1}{2} + \frac{1}{6} + \frac{1}{7} + \frac{1}{8} = \frac{157}{168},$$

which is not nearly competitive. If $b = 5$, then the sum is at most

$$\frac{1}{2} + \frac{1}{5} + \frac{1}{6} + \frac{1}{8} = \frac{119}{120},$$

nice but not our winner.

Now consider $b = 4$. In this case, if $c \geq 8$, then the best we can do is

$$\frac{1}{2} + \frac{1}{4} + \frac{1}{8} + \frac{1}{9} = \frac{71}{72}.$$

Thus, if $a = 2$ and $b = 4$, we need only consider the cases $c = 5$, 6, or 7. The optimal solutions for $c = 5$, 6, and 7, respectively, are

$$\frac{1}{2} + \frac{1}{4} + \frac{1}{5} + \frac{1}{21} = \frac{419}{420}, \quad \frac{1}{2} + \frac{1}{4} + \frac{1}{6} + \frac{1}{13} = \frac{155}{156}, \text{ and}$$

$$\frac{1}{2} + \frac{1}{4} + \frac{1}{7} + \frac{1}{10} = \frac{139}{140}.$$

None of these can match the sum (10.5). Thus, $a = 2$ and $b = 3$. We need to nail down c and d. The value of c must be less than 12, since otherwise the optimal sum becomes

$$\frac{1}{2} + \frac{1}{3} + \frac{1}{12} + \frac{1}{13} = \frac{155}{156}.$$

We need now consider the cases $c = 7$, 8, 9, 10, or 11, The optimal solutions for $c = 7$, 8, 9, 10, and 11, respectively, are

$$\frac{1}{2} + \frac{1}{3} + \frac{1}{7} + \frac{1}{43} = \frac{1805}{1806}, \quad \frac{1}{2} + \frac{1}{3} + \frac{1}{8} + \frac{1}{25} = \frac{599}{600},$$

$$\frac{1}{2} + \frac{1}{3} + \frac{1}{9} + \frac{1}{19} = \frac{341}{342},$$

$$\frac{1}{2} + \frac{1}{3} + \frac{1}{10} + \frac{1}{16} = \frac{239}{240}, \text{ and } \frac{1}{2} + \frac{1}{3} + \frac{1}{11} + \frac{1}{14} = \frac{230}{231}.$$

Of these, our first listed is best. Whew!

Please note that as the number of terms increases, each new solution simply adds on to the previous solution with identical initial summands. In fact, each solution involved the greedy algorithm where the denominators of successive terms was one more than the product of the previous terms. You can investigate further in the chapter Exercises.

Our final topic in this chapter is a variant of one of the most famous problems in all of mathematics. The ancient Greeks had determined all integral solutions to the equation

$$x^2 + y^2 = z^2, \tag{10.6}$$

the so-called Pythagorean triplets (since the solutions constitute the lengths of sides of a right triangle). It was determined that all *primitive* solutions (with x, y, and z pairwise relatively prime) are of the form $x = 2ab$, $y = a^2 - b^2$, and $z = a^2 + b^2$, where $a > b$, a and b are relatively prime and of opposite parity. Other solutions are just constant multiples of primitive solutions.

In 1637, when the French scholar and "amateur" mathematician Pierre de Fermat (1601–65) read about them in Diophantus's *Arithmetica*, he wrote directly in his copy of the book that in fact the equation

$$x^n + y^n = z^n \tag{10.7}$$

had no nontrivial integral solutions for any $n \geq 3$. He added, "I have discovered a truly marvelous proof, but the margin is too small to contain it." In fact, for centuries, no one was able to rediscover such a proof if he ever really had one and the conjecture became known as Fermat's Last Theorem. (I used to joke that it should have been called Fermat's Lost Theorem.) However, in 1994 a proof of the theorem was established after nearly a decade of concerted effort and some very deep insights by the British mathematician Andrew Wiles (then at Princeton University). We will not delve into such matters here. However, being a book on fractions, we ask for what values of n do there exist nontrivial integral solutions to

$$\frac{1}{x^n} + \frac{1}{y^n} = \frac{1}{z^n}, \tag{10.8}$$

which can be interpreted as a variant to equation (10.7) with negative exponents. In fact, this is another equation involving Egyptian fractions. The result is rather simple. I rediscovered it myself while in graduate school. The earliest mention of it that I am aware is due to J. D. Théron (1967).

Proposition 2: *Equation (10.8) is solvable for a given* n *if and only if equation (10.7) is solvable.*

By the Fermat–Wiles result for equation (10.7), equation (10.8) is solvable if and only if $n = 1$ or 2. Let's first establish the veracity of the proposition.

Proof: Let $x^n + y^n = z^n$ for some value of $n \geq 1$ and for positive integers $x, y,$ and z. Dividing by $x^n y^n z^n$, we obtain

$$\frac{1}{(yz)^n} + \frac{1}{(xz)^n} = \frac{1}{(xy)^n},$$

which is a solution to equation (10.8). Note, however, that if x, y, z form a primitive solution to equation (10.7) with x, y, z pairwise relatively prime, then the triple yz, xz, xy are no longer pairwise relatively prime (though they are still relatively prime in the sense that no integer greater than 1 divides all of them). Hence, a solution to equation (10.7) leads to a concomitant solution to equation (10.8).

Now suppose that

$$\frac{1}{r^n} + \frac{1}{s^n} = \frac{1}{t^n}.$$

for some value of $n \geq 1$ and for positive integers $r, s,$ and t. Multiplying by $r^n s^n t^n$, we obtain

$$(st)^n + (rt)^n = (rs)^n,$$

which is a solution to equation (10.7). Hence, for a given n, there is a one-to-one correspondence between solutions to equation (10.7) and those to equation (10.8).

In practice, this means that equations (10.7) and (10.8) have no non-trivial integral solutions for $n \geq 3$. However, there are solutions for $n = 1$ and $n = 2$. The case $n = 1$ is especially straightforward since from any solution to $x + y = z$, we get a corresponding solution to

$$\frac{1}{r} + \frac{1}{s} = \frac{1}{t}.$$

For example, the simple equation $1 + 2 = 3$ leads to $\frac{1}{6} + \frac{1}{3} = \frac{1}{2}$ where $6 = 2 \times 3$, $3 = 1 \times 3$, and $2 = 1 \times 2$. What sum of Egyptian fractions are created from $2 + 3 = 5$?

The case $n = 2$ is more interesting. Consider the Pythagorean triplet $3^2 + 4^2 = 5^2$. In this case, the corresponding solution to equation (10.8) with $n = 2$ is

$$\frac{1}{20^2} + \frac{1}{15^2} = \frac{1}{12^2}$$

since $20 = 4 \times 5$, $15 = 3 \times 5$, and $12 = 3 \times 4$. If instead we begin with our last fractional equation, we can create a solution to the Pythagorean equation (10.6) as in the proof of Proposition 2. In this case, we obtain $180^2 + 240^2 = 300^2$, where $180 = 15 \times 12$, $240 = 20 \times 12$, and $300 = 20 \times 15$. This last equation is not a primitive Pythagorean triplet. However, if we remove the factor of $\gcd(180, 240, 300) = 60$, we obtain our original primitive Pythagorean triplet $3^2 + 4^2 = 5^2$. Thus, there is a one-to-one correspondence between primitive Pythagorean triplets and solutions to the equation (10.6).

You can discover other examples in the following exercises.

Exercises

1. Write the following fractions as Egyptian fractions: $2/5$, $3/10$, $2/11$, $6/13$.
2. Write $7/12$ in two ways as the sum of Egyptian fractions.
3. Verify that the Erdös–Strauss equation

$$\frac{4}{n} = \frac{1}{x} + \frac{1}{y} + \frac{1}{z}$$

is solvable in positive integers x, y, and z (not necessarily distinct) for $2 \le n \le 13$.

4. Show that the equation

$$\frac{4}{n} = \frac{1}{x} + \frac{1}{y} + \frac{1}{z}$$

is solvable for all n of the form $4k + 3$ by confirming the algebraic identity

$$\frac{4}{4k + 3} = \frac{1}{k + 2} + \frac{1}{(k + 1)(k + 2)} + \frac{1}{(k + 1)(4k + 3)}.$$

5. Find distinct values for $n_1, n_2, \ldots, n_{10}$ for which $\frac{1}{n_1} + \frac{1}{n_2} + \ldots + \frac{1}{n_{10}}$ comes as close as possible to 2.

6. (a) Find a solution to

$$\frac{1}{x} + \frac{1}{y} = \frac{1}{z}.$$

   based on the equation $2 + 3 = 5$.

   (b) Find a solution to

$$\frac{1}{x} + \frac{1}{y} = \frac{1}{z}.$$

   based on the equation $3 + 4 = 7$.

7. (a) Find a solution to

$$\frac{1}{x^2} + \frac{1}{y^2} = \frac{1}{z^2}$$

   based on the Pythagorean triplet $5^2 + 12^2 = 13^2$.

   (b) Find a solution to

$$\frac{1}{x^2} + \frac{1}{y^2} = \frac{1}{z^2}$$

   based on the Pythagorean triplet $8^2 + 15^2 = 17^2$.

8. Find Pythagorean triplets using appropriate values of a and b and $x = 2ab$, $y = a^2 - b^2$, and $z = a^2 + b^2$.

9. What primitive Pythagorean triplet corresponds to the equation

$$\frac{1}{175^2} + \frac{1}{600^2} = \frac{1}{168^2}?$$

10. Show that if $p_1, p_2, \ldots, p_k$ are distinct primes, then no subset of the set

$$\left\{\frac{1}{p_1}, \frac{1}{p_2}, \ldots, \frac{1}{p_k}\right\}.$$

    sums to 1.

11. The Egyptians had a clever way for multiplying two integers (which has been imitated by many cultures since). Write the numbers m and n side by side. Below the number m, write half of m, then half again, and so on (ignoring any remainder) until the bottom entry is the number 1. Below the number n, calculate its double $2n$, then double that $4n$, and so on until there is a number beside each number in the left column. Add the numbers on the right that have odd numbers to their left. The sum will equal mn. Try this technique to calculate 72×59. Try it on 103×19. Pick your favorite product. Why does this work?

Chapter 11

Are All Real Numbers Rational?

> The art of detection is finding a common denominator for the fractions of a case.
>
> **Elsa Barker (1869–1954)**

We know that not all people are rational. But in this chapter, we are not interested in human nature, but rather in examining the content of the set of real numbers. Are all reals expressible as some sort of fraction of integers? If we pick up a stick and try to measure it based on some pre-determined unit length (inches, feet, meters, etc.), can we always build a ruler with fine enough markings to determine its length exactly? Though the stick might be between 26 and 27 inches, can we (at least in theory) create a ruler with markings of every $1/n$ of an inch for some n, so that we can measure it without error? Although there are infinitely many fractions (rational numbers) and they are "dense" in the sense that between any two of them lies infinitely more, it turns out the answer is "no."

This question was asked and unequivocally answered by the Pythagoreans in ancient Greece. In fact, something as simple as the diagonal of a unit square turns out to have irrational length. That is, if you try to measure the diagonal of a 1 foot by 1 foot square, it will be just a bit over 7/5 feet in length. If you have a ruler that measures in hundredths of a foot, it will be just a bit over 141/100 feet in length. No rational ruler, no matter how fancy, will give you an exact measurement. Of course, from the Pythagorean Theorem for right triangles, we know that the diagonal is of length $\sqrt{2}$ feet. So it suffices to show convincingly that $\sqrt{2}$ is not the same as some fraction. Here is the argument as recounted in Euclid's *Elements* more than 2300 years ago:

Suppose that $\sqrt{2}$ equals some fraction, say a/b. In fact, we insist that a/b be in reduced form so that a and b have no common factors larger than 1. We do this without loss of generality since all fractions can be so reduced. And now we have a fraction that is clearly and uniquely defined. Starting with $\sqrt{2} = a/b$, we square both sides of the equation to obtain $2 = a^2/b^2$. Next, multiply both sides by b^2 to get the equation

$$a^2 = 2b^2. \tag{11.1}$$

The right-hand side of equation (11.1) is even since it is divisible by 2. The same must be true of the left-side. So $2|a^2$. But if a were odd, then a^2 would be odd and not divisible by 2. Hence a itself is divisible by 2. Let's write $a = 2c$ for some integer c. Hence, equation (11.1) becomes $4c^2 = 2b^2$, and after division by 2, we have

$$b^2 = 2c^2. \tag{11.2}$$

But by the same reasoning, b^2 is even and so b is even as well. Thus, $2|b$ and hence $b = 2d$ for some integer d. But now both a and b are divisible by 2 and hence a and b are not relatively prime. This contradicts our assumption that we could write $\sqrt{2}$ as a fraction. Therefore, $\sqrt{2}$ is irrational.

By the way, here is another way to see that $\sqrt{2}$ is irrational which I consider even a bit simpler. Again, suppose otherwise. Let $\sqrt{2} = a/b$ for some fraction a/b reduced or not. Square both sides and simplify to obtain formula (11.1) once again: $a^2 = 2b^2$. By the Fundamental Theorem of Arithmetic, both a and b have unique prime factorizations (a fact that we subtly assumed in our previous proof when we required a and b to be unique). No matter how many factors of 2 divide b, the number b^2 will have double that number. Thus, the right-hand side of equation (11.1) is a number with an odd number of factors of 2, namely one more than the even number of 2's in the prime factorization of b^2. But by similar reasoning, a^2 must have an even number of 2's in its prime factorization. So it can't be the case that $a^2 = 2b^2$. Hence, we've shown again that $\sqrt{2}$ is irrational.

You might find it amusing to extend our result by showing that $\sqrt{n}$ is irrational as long as n is not a perfect square. For that matter, $\sqrt[m]{n}$ is irrational for any n not a perfect mth power. Similarly, we can even show that numbers like $\log_a b$ are irrational if b is not a rational power of a. And it can be shown that most real numbers are actually irrational in a carefully defined mathematical sense. The rational numbers that we study in this book are *countable*, whereas the irrational numbers are *uncountable*. If you had an infinitely sharp dart and could throw it at a real number line, you would have a 100% chance of hitting an irrational number rather than a rational number. That's not to say that our study of fractions is much ado about nothing. It's not impossible to hit a rational number, you just have less than any positive probability of doing so.

We won't delve very deeply into the study of the cardinality of numbers beautifully developed by the great German mathematician Georg Cantor. Even Cantor himself may have been influenced by Galileo (1564–1642), who noticed that it could be argued there were as many positive square numbers as there were positive integers (natural numbers) generally. On the one hand, of course all perfect squares 1, 4, 9, 16, etc. are positive integers and yet there are an infinite number of positive integers left over that are not perfect squares: 2, 3, 5, 6, etc. So apparently there are many more positive integers than there are squares. But, on the on the other hand, for each positive integer we can perfectly match it up with its square: $1 \leftrightarrow 1$, $2 \leftrightarrow 4$, $3 \leftrightarrow 9$, $4 \leftrightarrow 16$, etc. so that each positive integer has a matching square and vice versa. The key idea that if two sets can be matched up via such a one-to-one correspondence, then they have the same *cardinality*; this goes back at least to Galileo. And this concept of cardinality applies to infinite as well as to finite sets. The "smallest" infinite sets are those that are countable (or listable) such as the set of natural numbers.

Let us see why the set of all fractions is countable in the sense that we can delineate them in a specified list. Our list will be infinitely long, but we can be certain that any particular rational number is included on it. In particular, we will create a list of all the positive rational numbers. Our algorithm (or recipe) for writing out the list will actually include all forms of all the rationals reduced and otherwise. Indeed our sequence will actually list each and every positive fraction an infinite number of times. But at least it will include them all in a nice orderly fashion. Here's how it works:

Start by listing all the fractions a/b with sum $a + b$ equaling 2. There is just one such fraction, namely 1/1. Next list those with $a + b = 3$, then $a + b = 4$, etc. For each value of the sum, begin with the smallest numerator (namely 1) and follow by adding 1 to the next numerator while subtracting 1 from the denominator until the denominator is as small as possible (namely 1). Hence, the sequence begins

1/1, 1/2, 2/1, 1/3, 2/2, 3/1, 1/4, 2/3, 3/2, 4/1, 1/5, 2/4, 3/3,

4/2, 5/1, 1/6, etc.

This way no fraction is missed. Each new group of fractions begins with $1/n$ and ends with $n/1$. All other fractions of the form a/b appear in the group having sum $a + b$ sitting between $(a - 1)/(b + 1)$ and $(a + 1)/(b - 1)$. For example, 3/7 sits in the 10's group directly between 2/8 and 4/6.

We can even specify the exact position of each of our fractions. Recall that the sum of the integers from 1 to n is $n(n + 1)/2$. This is known as the nth triangular number since we can arrange the sum into a set of dots forming a pleasant isosceles triangle (Figure 11.1). Restated,

$$1 + 2 + 3 + \ldots + n = \frac{n(n + 1)}{2}. \tag{11.3}$$

Consider the fraction a/b. It is preceded by all the fractions having numerator plus denominator sum less than $a + b$ together with all the fractions of sum $a + b$ having numerator less than a. There is one fraction with sum 2 (namely 1/1), two fractions with sum 3 (namely 1/2 and 2/1), three fractions with sum 4 (namely, 1/3, 2/2, and 3/1), ..., and $a + b - 2$ fractions with sum $a + b - 1$. The number of fractions with numerator plus denominator sum less than $a + b$ is thus

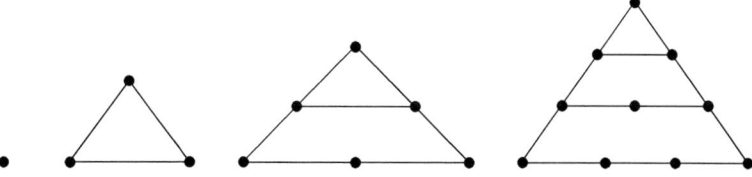

Figure 11.1 First four triangular numbers.

$$\frac{(a + b - 2)(a + b - 1)}{2},$$

obtained by letting $n = a + b - 2$ in formula (11.3). The fraction a/b is the ath fraction in the next group. We can now specify the precise position of a/b in our list. The fraction a/b sits in position

$$\frac{(a + b - 2)(a + b - 1)}{2} + a. \tag{11.4}$$

Let's illustrate formula (11.4) with an example. For the fraction 5/2, $a = 5$, $b = 2$. So the fraction 5/2 should appear in position

$$\frac{5 \cdot 6}{2} + 5 = 20.$$

Let's double check by counting along: 1/1, 1/2, 2/1, 1/3, 2/2, 3/1, 1/4, 2/3, 3/2, 4/1, 1/5, 2/4, 3/3, 4/2, 5/1, 1/6, 2/5, 3/4, 4/3, 5/2, ... Did you verify by counting?

Exercises

1. (a) Show that $\sqrt{3}$ is irrational.
   (b) Why doesn't your proof work on $\sqrt{4}$?
2. Show that $\log_{10} 2$ is irrational.
3. Show that the sum of the first n odd numbers equals n^2.
4. Verify that 21, 2211, 222, 111, 22, 221, 111, . . . are all triangular numbers.
5. Verify that 41,616 is both a triangular number and a square. Euler (1730) proved that there is an infinity of such examples.

6. Find a formula for the nth pentagonal number $p_n = 1 + 4 + 7 + 10 + \ldots + (3n - 2)$.

7. (a) In our list of rationals, which position is held by 3/5? How about 10/7?

   (b) What number appears in position 1000?

8. Consider the expression

$$\left(\sqrt{2}^{\sqrt{2}}\right)^{\sqrt{2}}$$

to show that an irrational raised to an irrational power can be a rational number.

9. Is the number 0.12345678910111213 ... consisting of concatenating all the positive integers in order rational or irrational?

Chapter 12

Continued Fractions

> The Bible tells man how to go to heaven, not how the heavens go.
>
> **Galileo Galilei (1564–1642)**

We now look at a fascinating area of mathematics involving fractions within fractions aptly called *continued fractions*. Such expressions have the form

$$a_0 + \cfrac{b_1}{a_1 + \cfrac{b_2}{a_2 + \cfrac{b_3}{a_3 + \cdots}}},$$

where the a_i's and b_j's are all integers. If all the b_j's are equal to 1 and $a_i \geq 1$ for all $i \geq 1$, then it is a *simple continued fraction*. With all 1s in the numerators, notice how the spirit of Egyptian fractions is retained in simple continued fractions. Furthermore, if the sum terminates rather than going on ad infinitum, then the sum is a *finite simple continued fraction*. In this chapter, we will be mainly focused on these latter fractions, but in later chapters we will address some special cases of other continued fractions.

For the sake of notational simplicity, the simple continued fraction

$$a_0 + \cfrac{1}{a_1 + \cfrac{1}{a_2 + \cfrac{1}{a_3 + \cdots}}}$$

will henceforth be denoted simply by $[a_0; a_1, a_2, a_3, \ldots]$. The a_i are called the *partial quotients* of the associated continued fraction. Shortly, we'll see why this is an apt moniker for them. But first, let's look at an easy example.

Let's determine what rational number is represented by the simple continued fraction $r = [1; 2, 3, 4, 5]$. Recall this is an abbreviated notation for the fraction

$$1 + \cfrac{1}{2 + \cfrac{1}{3 + \cfrac{1}{4 + \frac{1}{5}}}}.$$

To simplify, work from the bottom up. The mixed fraction $4 + \frac{1}{5}$ is equal to the (so-called) improper fraction $\frac{21}{5}$. Its reciprocal is $\frac{5}{21}$. Hence,

$$r = 1 + \cfrac{1}{2 + \cfrac{1}{3 + \frac{5}{21}}}.$$

But $3 + \frac{5}{21} = \frac{68}{21}$. Its reciprocal is $\frac{21}{68}$. Thus,

$$r = 1 + \cfrac{1}{2 + \frac{21}{68}}.$$

Next, $2 + \frac{21}{68} = \frac{157}{68}$ with reciprocal $\frac{68}{157}$. Finally,

$$r = 1 + \frac{68}{157} = \frac{225}{157}.$$

The process is straightforward, though it could become a bit tedious if there is a long list of partial quotients. For practice, see if you can simplify [3; 1, 4, 1, 5]. You should verify that the answer is 134/35. In this chapter, a much faster method of simplifying a given simple continued fraction will be presented.

The study of continued fractions has a long and interesting history dating back at least to the ancient Greeks, though they didn't present their work in this form explicitly. In addition, continued fractions have found application to such diverse areas as differential equations, statistics, operations research, electrical networks, and even astronomy.

Of course, any finite simple continued fraction will represent some rational number. The converse is also true as we demonstrate presently.

Proposition: *Any rational number can be expressed as a finite simple continued fraction.*

Proof: Let $r = a/b$ be an arbitrary rational number and suppose that

$$a/b = a_0 + r_1/b,$$

where a_0 is the greatest integer less than or equal to r. If r is an integer, then $r = a_0$ and we are done. If r is not an integer, then $r_1 \neq 0$ and $1 \leq r_1 < b$ since $r_1/b < 1$. Hence,

$$b/r_1 = a_1 + r_2/r_1,$$

where a_1 is the greatest integer less than or equal to b/r_1. It necessarily follows that $0 \leq r_2 < r_1$. If $r_2 = 0$, then we are done. Otherwise, write

$$r_1/r_2 = a_2 + r_3/r_2,$$

where $0 \leq r_3 < r_2$, and so on. In general, the ith equation has the form

$$r_{i-2}/r_{i-1} = a_{i-1} + r/r_{i-1}.$$

Since the r_i's form a decreasing sequence of nonnegative integers, there must be an n for which $r_n > 0$ while $r_{n+1} = 0$. Working backward,

$$r_{n-2}/r_{n-1} = a_{n-1} + 1/a_n = [a_{n-1}; a_n]$$

$$r_{n-3} = a_{n-2} + \frac{1}{r_{n-2}/r_{n-1}} = [a_{n-2}; a_{n-1}, a_n]$$

.

.

.

and finally $r = [a_0; a_1, \ldots, a_n]$. This completes our proof.

Notice that if we multiply the ith equation through by r_{i-1} for all i, then we get

$$r_{i-2} = a_{i-1}r_{i-1} + r_i,$$

where $0 \leq r_i < r_{i-1}$. This is identical to the Euclidean algorithm for determining the greatest common divisor of a and b previously studied. The quotients q_i that appear in the Euclidean algorithm have simply been replaced by the a_i above. It should now be clear why the a_i are called partial quotients.

Example of a finite simple continued fraction: Let's express 109/25 as a finite simple continued fraction.

Solution:

$$109 = 4 \times 25 + 9$$
$$25 = 2 \times 9 + 7$$
$$9 = 1 \times 7 + 2$$
$$7 = 3 \times 2 + 1$$
$$2 = 2 \times 1$$

Hence, it follows that $109/25 = [4; 2, 1, 3, 2]$.

We can check directly,

$$\frac{109}{25} = 4 + \frac{9}{25} = 4 + \frac{1}{\frac{25}{9}}$$

$$= 4 + \cfrac{1}{2 + \frac{7}{9}} = 4 + \cfrac{1}{2 + \frac{1}{9/7}}$$

$$= 4 + \cfrac{1}{2 + \cfrac{1}{1 + \frac{2}{7}}} = 4 + \cfrac{1}{2 + \cfrac{1}{1 + \frac{1}{7/2}}}$$

$$= 4 + \cfrac{1}{2 + \cfrac{1}{1 + \cfrac{1}{3 + \frac{1}{2}}}}$$

which is the same as $[4; 2,1,3,2]$. Since $2 = 1 + 1$, we can write $109/25 = [4; 2, 1, 3, 1, 1]$ as well. So, there are two ways to write the fraction 109/25 as a finite simple continued fraction. Generally, any simple continued fraction with final partial quotient an integer $a > 1$ is equivalent to a fraction with a replaced by $a - 1$ with a 1 appended after it. Analogously, a simple continued fraction ending with final partial quotient 1 is equivalent to the same fraction with 1 removed and 1 added to the partial quotient preceding it. If we want to retain uniqueness for consistency's sake, we can simply always insist that the last partial quotient not be a 1. Note that this is similar to the situation with decimal expansions where any terminating decimal could be replaced with a nonterminating decimal. For example, $2.2 = 2.1999\bar{9}$.

Let's look back at our example, the simple continued fraction [4: 2, 1, 3, 2]. Let's see what happens as we build up to the full fraction. In this case, $[4] = 4 = 4.00$, $[4; 2] = \frac{9}{2} = 4.50$, $[4 : 2, 1] = \frac{13}{3} = 4.3\bar{3}...$, $[4; 2, 1, 3] = \frac{48}{11} = 4.36\overline{36}$, and finally $[4: 2, 1, 3, 2] = 109/25 = 4.36$. Notice that the preliminary sums bounce back and forth between being larger and smaller than the final target, yet get progressively closer in absolute value. This is a general phenomenon. To be able to speak more concretely, we make the following definition:

Let $r = [a_0; a_1, ..., a_n]$ and let $r_i = [a_0; a_1, .., a_i]$ where $0 \leq i \leq n$. Then the rational number r_i is called the *ith convergent to r*.

In our previous example, the convergents to 109/25 were 4, 9/2, 13/3, 48/11, and 109/25, respectively. Each is a reduced fraction getting progressively closer to the final answer. Interestingly, there is a handy recursive technique for readily calculating all the convergents. We let p_i be the numerator of the ith convergent and let q_i denote its denominator. To get the process going, we need to "seed" a couple initial values for p_i and q_i as well. Here is the result: Let $r = [a_0; a_1, ..., a_n]$ and let $r_i = \frac{p_i}{q_i}$ be the ith convergent to r for $0 \leq i \leq n$. Let $p_{-2} = q_{-1} = 0$ and $p_{-1} = q_{-2} = 1$. Then

$$p_i = a_i p_{i-1} + p_{i-2} \text{ and } q_i = a_i q_{i-1} + q_{i-2} \text{ for } 0 \leq i \leq n. \quad (12.1)$$

Example: Let $r = [1; 2, 3, 4, 5]$. To get things rolling, we enter all the a_i's and p_i and q_i for negative i into our table (in bold). Then formula (12.1) allows us to quickly calculate successive values of the p's and q's.

i	-2	-1	0	1	2	3	4
a_i			1	2	3	4	5
p_i	0	1	1	3	10	43	225
q_i	1	0	1	2	7	30	157

Thus, the successive convergents are $r_0 = 1$, $r_1 = 3/2$, $r_2 = 10/7$, $r_3 = 43/30$, and $r_4 = r = 225/157$.

Calculating the differences $r - r_i$ for $0 \leq i \leq 3$ (to six decimal places' accuracy) yields

$$225/157 - 1 = 68/157 = 0.433121\ldots$$
$$225/157 - 3/2 = -21/314 = -0.066879\ldots$$
$$225/157 - 10/7 = 5/1099 = 0.004550\ldots$$
$$225/157 - 43/30 = -0.000212\ldots$$

Again, the convergents alternate back and forth between being larger then smaller than r. Here

$$r_0 < r_2 < r_4 = r < r_3 < r_1.$$

This series of inequalities holds generally although it would be a lengthy sidetrack to prove the result in this book. But let's at least establish why formula (12.1) must always be true. We proceed with our old friend, mathematical induction.

Proof of (12.1) By Induction: For $i = 0$, $r_0 = a_0/1$, and hence

$$p_0 = a_0 = a_0 \cdot 1 + 0 = a_0 p_{-1} + p_{-2} \text{ and } q_0 = 1 = a_0 \cdot 0 + 1$$
$$= a_0 q_{-1} + q_{-2}.$$

Now suppose the result is true for $i = k$ for some k where $0 \leq k < n$. Then

$$r_{k+1} = \frac{p_{k+1}}{q_{k+1}} = [a_0; a_1, \ldots, a_k, a_{k+1}] = [a_0; a_1, \ldots, a_{k-1}, a_k + 1/a_{k+1}].$$

The latter continued fraction has the same number of partial quotients as the former, although its last partial quotient no longer need be an integer. Applying the inductive hypothesis to the latter continued fraction, we obtain

$$p_{k+1} = (a_k + 1/a_{k+1}) p_{k-1} + p_{k-2} \text{ and } q_{k+1} = (a_k + 1/a_{k+1}) q_{k-1} + q_{k-2}.$$

It follows that

$$
\begin{aligned}
r_{k+1} &= \frac{\left(a_k + \frac{1}{a_{k+1}}\right) p_{k-1} + p_{k-2}}{\left(a_k + \frac{1}{a_{k+1}}\right) q_{k-1} + q_{k-2}} \\
&= \frac{(a_k \cdot a_{k+1} + 1) p_{k-1} + a_{k+1} \cdot p_{k-2}}{(a_k \cdot a_{k+1} + 1) q_{k-1} + a_{k+1} \cdot q_{k-2}} \\
&= \frac{a_{k+1} (a_k p_{k-1} + p_{k-2}) + p_{k-1}}{a_{k+1} (a_k q_{k-1} + q_{k-2}) + q_{k-1}} \\
&= \frac{a_{k+1} p_k + p_{k-1}}{a_{k+1} q_k + q_{k-1}}.
\end{aligned}
$$

Hence, $p_{k+1} = a_{k+1} p_k + p_{k-1}$ and $q_{k+1} = a_{k+1} q_k + q_{k-1}$. The case $i = k + 1$ follows and our formula is confirmed.

Next, we note an additional captivating property of convergents. If we look back at our chart for $r = [1; 2, 3, 4, 5]$ and calculate $p_i q_{i-1} - p_{i-1} q_i$ for $1 \leq i \leq 4$, the result will be either 1 or -1. In fact, if $r = [a_0; a_1, \ldots, a_n]$ and $r_i = p_i/q_i$ is its ith convergent, then

$$p_i q_{i-1} - p_{i-1} q_i = (-1)^{i-1} \text{ for } 0 \leq i \leq n. \tag{12.2}$$

Formula (12.2) can be easily established via induction. For $i = 0$, $p_0 q_{-1} - p_1 q_0 = a_0 \cdot 0 - 1.1 = (-1)^{0-1}$. Next assume the assertion is true for $i = k$ for some k with $0 \leq k < n$.

We show that the case $i = k + 1$ necessarily follows. By formula (12.1)

$$
\begin{aligned}
p_{k+1} q_k - p_k q_{k+1} &= (a_{k+1} p_k + p_{k-1}) q_k - p_k (a_{k+1} q_k + q_{k-1}) \\
&= -(p_k q_{k-1} - p_{k-1} q_k) = -(-1)^{k-1} = (-1)^k.
\end{aligned}
$$

Thus, the case $k + 1$ is established which completes the proof by induction.

As a corollary to formula (12.2), we offer the striking result.

If r_i is the ith convergent to $r = [a_0; a_1, \ldots, a_n]$, then for any i with $1 \le i \le n$,

$$r_i - r_{i-1} = \frac{(-1)^{i-1}}{q_i q_{i-1}}. \tag{12.3}$$

Verify this in our example for various i. For example, for $i = 3$, $r_3 - r_2 = \frac{43}{30} - \frac{10}{7} = \frac{1}{210}$.

Here's why formula (12.3) is a corollary to formula (12.2). For $1 \le i \le n$,

$$r_i - r_{i-1} = \frac{p_i}{q_i} - \frac{p_{i-1}}{q_{i-1}} = \frac{p_i q_{i-1} - p_{i-1} q_i}{q_i q_{i-1}} = \frac{(-1)^{i-1}}{q_i q_{i-1}}.$$

Since the q's keep getting larger,

$$|r_i - r_{i-1}| < \frac{1}{q_{i-1}^2}.$$

And since the partial quotients alternate between being larger and smaller than the target r, it follows that

$$|r - r_i| < \frac{1}{q_i^2} \text{ for all } i \ge 1.$$

Hence, we see that the partial quotients of r are excellent approximations to it. Though we won't prove it here, it turns out that they are the *best* rational approximations to r in a concrete sense: If $r = [a_0; a_1, \ldots, a_n]$ with convergents $r_i = \frac{p_i}{q_i}$ for $0 \le i \le n$, then if some fraction $\frac{p}{q}$ satisfies the inequality

$$\left| r - \frac{p}{q} \right| < |r - r_i|,$$

it must be the case that $q > q_i$. In other words, any better approximation to r must have a larger denominator. Equivalently, the convergents are the best approximations to r among all fractions with the same or smaller denominators.

And all of this can be taken to the next level concerning *infinite* simple continued fractions!

If $[a_0; a_1, \ldots]$ is an infinite simple continued fraction with convergents r_n for $n \geq 0$, then there is a real number $x = \lim_{n \to \infty} r_n$. Unlike finite simple continued fractions, the real number x won't be a rational number any longer but rather some irrational real number. We won't give the technical proof that the limit exists here, but hopefully it seems highly reasonable.

Luckily, what we've discovered about finite simple continued fractions for rational r generalizes directly in the infinite case with any real x. The convergents can be calculated via the same algorithm that we studied, the convergents ping pong back and forth homing in on x, and each convergent is the best approximation to x with denominators of that size or less. Furthermore, quadratic surds (as they are called) like $\sqrt{2}$ or $\sqrt{5}$ have surprisingly regular continued fraction expansions. A quadratic surd is an irrational number that is the root of a quadratic polynomial with rational coefficients. Amazingly, they all have a repeating pattern!

To take a good example, let's find the infinite simple continued fraction for $x = \sqrt{2}$. A little algebraic manipulation is all that's needed. Notice that $(\sqrt{2} + 1)(\sqrt{2} - 1) = 1$. Thus,

$$\sqrt{2} - 1 = \frac{1}{1 + \sqrt{2}}$$

and

$$\sqrt{2} = 1 + \frac{1}{1 + \sqrt{2}}.$$

Next, "plug in" the above expression for $\sqrt{2}$ for the $\sqrt{2}$ sitting in the denominator on the right:

$$\sqrt{2} = 1 + \cfrac{1}{1 + 1 + \cfrac{1}{1 + \sqrt{2}}} = 1 + \cfrac{1}{2 + \cfrac{1}{1 + \sqrt{2}}}.$$

Substituting again

$$\sqrt{2} = 1 + \cfrac{1}{2 + \cfrac{1}{1+\sqrt{2}}} = 1 + \cfrac{1}{2 + \cfrac{1}{1+1+\cfrac{1}{1+\sqrt{2}}}} = 1 + \cfrac{1}{2 + \cfrac{1}{2+\cfrac{1}{1+\sqrt{2}}}}.$$

Continuing ad infinitum, $\sqrt{2} = [1; 2, 2, 2, \ldots]$ where thankfully we make use of our more compact notation. Next, we find an approximation to $\sqrt{2}$ accurate to at least two decimal places. Begin with our usual table to calculate successive convergents. Again, the bold-faced numbers are our given values for the partial quotients, while the rest of the numbers are calculated using formula (12.1). We'll fill out the table until the q's surpass 100.

I	-2	-1	0	1	2	3	4	5	6	7
a_i			**1**	**2**	**2**	**2**	**2**	**2**	**2**	**2**
p_i	0	1	1	3	7	17	41	99	239	577
q_i	1	0	1	2	5	12	29	70	169	408

It follows that the fraction $\frac{577}{408}$ is the best rational approximation to $\sqrt{2}$ with denominator at most 408. In fact, $\sqrt{2} - \frac{577}{408} = 0.0000021\ldots$, a much better approximation than even hoped for.

By a theorem established by the eminent French mathematician J. L. Lagrange (1736–1814), all quadratic surds will also have periodic continued fractions, though the algebra can get a bit more tangled. In our next interlude, we'll work out the continued fraction for $\sqrt{3}$. Meanwhile, let's take a look at the simplest looking infinite continued fraction $[1; 1, 1, 1, \ldots]$. Can we determine what real number x it represents?

In this case,

$$x = 1 + \cfrac{1}{1 + \cfrac{1}{1+\cfrac{1}{1\ldots}}} = 1 + \cfrac{1}{x}.$$

Hence, $x^2 = x+1$. Look familiar? Solving with the quadratic formula, we get

$$x = \frac{1 \pm \sqrt{5}}{2}.$$

But x must be positive and so

$$x = \frac{1 + \sqrt{5}}{2}.$$

This is our old friend φ. Let's take a look at its convergents.

i	-2	-1	0	1	2	3	4	5	6	7	8	9
a_i			1	1	1	1	1	1	1	1	1	1
p_i	0	1	1	2	3	5	8	13	21	34	55	89
q_i	1	0	1	1	2	3	5	8	13	21	34	55

Yes, the convergents are all ratios of successive Fibonacci numbers. We've already seen that the limit of their ratios is the golden ratio φ. But now we see that the ratios of Fibonacci numbers are the *best* rational approximations to φ!

At this point, you may wonder about continued fractions for other real numbers. Not surprisingly, greater analysis is required. Let x be any irrational number. To get the process started, let $x = A_0$. Next, let $a_0 = [A_0]$, the greatest integer less than or equal to A_0. Then let

$$A_1 = \frac{1}{A_0 - a_0} \text{ and } a_1 = [A_1].$$

Similarly, define a_n recursively for all $n \geq 1$ by

$$A_n = \frac{1}{A_{n-1} - a_{n-1}} \text{ and } a_n = [A_n]. \tag{12.4}$$

Since A_0 is irrational, A_1 is irrational. But then so is A_2, A_3, A_4 and so on. Hence, the sequence $a_0, a_1, a_2, a_3, a_4, \ldots$ never terminates. Furthermore, $A_0 - a_0 < 1$ implies that $A_1 > 1$ and analogously (from formula (12.4)), $A_n > 1$ implies that $a_n \geq 1$ for all $n \geq 1$.

It follows from formula (12.4) that

$$x = a_0 + \frac{1}{A_1} = a_0 + \frac{1}{a_1 + \frac{1}{A_2}} = \ldots = [a_0; a_1, a_2, \ldots, a_n, A_{n+1}] \text{ for all } n \geq 1.$$

$$(12.5)$$

In this instance, note that all the a's are integers, but that A_{n+1} is not even rational. Even so, the first n convergents

$$r_i = \frac{p_i}{q_i}$$

are rational for $0 \leq i \leq n$ while

$$x = \frac{p_{n+1}}{q_{n+1}}$$

in formula (12.5).

But as n gets arbitrary large, the difference between x and r_n gets arbitrarily small. Hence, $\lim_{n \to \infty} |x - r_n| = 0$ and so $x = [a_0; a_1, a_2, \ldots]$.

By utilizing equation (12.4), we can determine the initial part of the continued fraction for any real number if we have some information about its decimal expansion. Let's take a look at the constant $\pi = 3.14159265\ldots$, the ratio of the circumference to the diameter of any circle. We begin with $A_0 = \pi$ and then successively calculate the subsequent a's and A's:

$$A_0 = \pi = 3.14159265\ldots \quad a_0 = [A_0] = 3$$
$$A_1 = 1/(A_0 - a_0) = 7.06251331\ldots \quad a_1 = [A_1] = 7$$
$$A_2 = 1/(A_1 - a_1) = 15.99659409\ldots \quad a_1 = [A_2] = 15$$
$$A_3 = 1/(A_2 - a_2) = 1.00341722\ldots \quad a_3 = [A_3] = 1$$
$$A_4 = 1/(A_3 - a_3) = 292.63483365\ldots \quad a_4 = [A_4] = 292$$
$$A_5 = 1/(A_4 - a_4) = 1.57521580\ldots \quad a_5 = [A_5] = 1$$
$$A_6 = 1/(A_5 - a_5) = 1.73847795\ldots \quad a_6 = [A_6] = 1$$

Thus, $\pi = [3; 7, 15, 1, 292, 1, 1, \ldots]$. We can tabulate its initial convergents.

i	-2	-1	0	1	2	3	4	5	6
a_i			3	7	15	1	292	1	1
p_i	0	1	3	22	333	355	103,933	104,348	208,341
q_i	1	0	1	7	106	113	33,102	33,215	66,317

The greatest mathematician of antiquity is no doubt Archimedes of Syracuse (287–212 BCE) He was able to demonstrate that $220/71 < \pi < 22/7$. Notice that his upper bound was one of the convergents to π found above. Coincidence? We'll say a bit more about this in Chapter 13. Interestingly, the Chinese mathematician Tsu Chung-Chi (430–501 CE) explicitly described $355/113$ as being an excellent approximation to π. This approximation is especially easy to remember by simply writing 113,355 and separating the number in the center.

The convergents to π converge rather quickly. By way of comparison, note that

$$\pi - 208,341/66,317 = 0.0000000001223....$$

We complete this chapter with a brief discussion of an application of continued fractions in making a workable planetarium. The first planetaria were constructed in ancient Greece, including one built by Archimedes. Over time, gradually improvements were made. The Dutch scientist and mathematician Christiaan Huygens (1629–95) was determined to make one of even greater accuracy. Huygens was already famous for his mathematical works, his invention of a pendulum clock using an inverted cycloidal arch, his modern theories of light, and his observation of the rings of Saturn (see Figure 12.1).

In 1680–1, Huygens designed and built a cogwheeled planetarium (often called an orrery) at the behest of the French l'Academie Royale des Sciences. To do so, he utilized the published observations of the insightful but idiosyncratic astronomer Johann Kepler (actually based somewhat on observations of the Danish astronomer Tycho Brahe). The planetarium included all the known planets at the time—Mercury, Venus, Earth, Mars, Jupiter, and Saturn. The

Figure 12.1 Christiaan Huygens, painted by Bernard Vaillant, 1686.

Reproduced from Huygensmuseum Hofwijck, Voorburg via Wikimedia Commons.

planets ran along circular paths, but he placed the sun a bit off-center to at least suggest elliptical orbits. One full rotation of the handle would move the planets ahead one Earth year. A reverse rotation would move the planets back a year. The planetarium itself used a complex set of rotating gears, each with a varying number of teeth. Amazingly, Huygens made careful use of continued fractions!

Let's take a look at the calculations involved with Saturn. Huygens used the observations that in 365 days, the Earth covers $359°45'40''31'''$ and Saturn $12°13'34''18'''$ of its orbit. Notice our modern use of sexagesimal fractions! Their ratio is given by

$$\frac{359 + \frac{45}{60} + \frac{40}{3,600} + \frac{31}{216,000}}{12 + \frac{13}{60} + \frac{34}{3,600} + \frac{18}{216,000}} = \frac{77,708,431}{2,640,858}$$

Working out the continued fraction expansion for the above fraction using equation (12.4) yields

$$\frac{77,708,431}{2,640,858} = [29; 2, 2, 1, 5, 1, 4, 1, 1, 2, 1, 6, 1, 10, 2, 2, 3]$$

$$= 29.425448. \ldots$$

Next, we utilize equation (12.1) to determine the first few numerators and denominators of the convergents.

i	−2	−1	0	1	2	3	4	5
a_i			29	2	2	1	5	1
p_i	0	1	29	59	147	206	1177	1383
q_i	1	0	1	2	5	7	40	47

Of course, Huygens was especially bound by engineering constraints and machinist's limitations. Hence, he chose a gear ratio of $206/7 = 29.\overline{428571}$ for the planetarium based on the third convergent above. This gives an approximation error of

$$\left| \frac{77708431}{2640858} - \frac{206}{7} \right| = 0.003123\ldots.$$

For Mercury, Huygens used other observations, leading to the ratio of

$$\frac{25,335}{105,190} = [0; 4, 6, 1, 1, 2, 1, 1, 1, 1, 7, 1, 2].$$

The appropriate convergents table becomes

i	−2	−1	0	1	2	3	4	5	6	7	8	9
a_i			0	4	6	1	1	2	1	1	1	1
p_i	0	1	0	1	6	7	13	33	46	79	125	204
q_i	1	0	1	4	25	29	54	137	191	328	519	847

Huygens could have used the third convergent of 7/29 as gear ratios and been quite satisfied. But he wished to do better. Initially, he planned to use the fifth convergent of 33/137 but then hit upon the idea of using four interlocking gears for a pair of planets rather than two. He used the ninth convergent! By factoring $204 = 12 \times 17$ and $847 = 7 \times 121$, a gear with 121 teeth was connected to one with 12 teeth and a gear with 7 teeth with one with 17. All told, his planetarium worked spectacularly well for Mercury's path. The error of approximation is just

$$\left| \frac{25335}{105190} - \frac{204}{847} \right| = 0.000000168.\ldots$$

Quite an accomplishment!

Exercises

1. Determine the simple continued fraction for the following:
   (a) 100/17
   (b) 5/101
   (c) 89/144.
2. Determine the fraction represented by the continued fraction [1; 2, 2, 2, 2].
3. A tropical year on Earth is approximately 365.2422 days. Determine the simple continued fraction for 365.2422.
4. Determine the first few convergents for Earth's tropical year.
5. Verify that the continued fraction [5; 5, 10, 10] is a pretty good approximation to $3\sqrt{3}$.
6. Find the simple continued fraction expansion for $\sqrt{7}$.
7. Determine the simple continued fractions for the following quadratic surds:
   (a) $\sqrt{5}$
   (b) $\sqrt{10}$
   (c) $\sqrt{17}$
   (d) $\sqrt{k^2 + 1}$.
8. Show that the golden ratio φ is also equal to
$$\sqrt{1 + \sqrt{1 + \sqrt{1 + \ldots}}}$$
9. If x is any irrational number and q is any positive integer, then there is rational number p/q with
$$\frac{p}{q} - \frac{1}{2q} < x < \frac{p}{q} + \frac{1}{2q}.$$

10. For the inequality in problem 9, find p for the given x and q:
   (a) $x = \sqrt{2}$, $q = 10$
   (b) $x = \sqrt{3}$, $q = 8$
   (c) $x = \pi$, $q = 7$
   (d) $x = \pi$, $q = 12$.

Chapter 13
Archimedes' Determination of Pi

> Mathematics reveals its secrets only to those who appreciate it with pure love, for its own beauty.
>
> **Archimedes (*c*.287–212 BCE)**

Archimedes of Syracuse (287–212 BCE) was the greatest scientist and geometer of ancient times (Figure 13.1). In fact, he is usually considered to be one of the three or four greatest mathematicians of all time. As Voltaire said, "there was more imagination in the head of Archimedes than in that of Homer."

Figure 13.1 Archimedes, oil painting by Domenico Fetti, 1620.
Reproduced from Gemäldegalerie Alte Meister via *Wikimedia Commons.*

To the general public, he is probably best known for the stunning military defense of Syracuse against the invading Roman army that lasted for more than two years. To do so, he built various catapults, grappling cranes to lift ships out of the harbor, and perhaps even a giant concave mirror that supposedly was used to set the Roman ships on fire. He was a consummate physicist and engineer. He invented the Archimedean screw to raise water for irrigation and made a crude odometer that could be attached to a rolling cart to measure distance. He also wrote *On Floating Bodies*, a work on hydrostatics that included the law that a body immersed in a fluid is buoyed up by a force equal to the weight of the displaced fluid. (This is the genesis of the story of him running naked and shouting "Eureka!" when he was able to prove that the king's crown was not all gold.) He also wrote *On Levers*, detailing the science of simple machines such as pulleys, inclined planes, levers, etc., as well as a treatise *Plane Equilibriums* which contains twenty-five propositions on *statics*, including the centers of gravity of various plane laminas. Despite all of this, like Plato (427–347 BCE), he felt that such practical matters were below the dignity of a philosopher.

In matters of pure mathematics, Archimedes stands without peer in the ancient world. Though much has been lost to time, two works on solid geometry are considered masterpieces even today. *The Sphere and the Cylinder* contains fifty-three propositions in two books where he derives formulas for volumes and surface areas of pyramids, cones, spheres, and for surfaces of revolution of inscribed polygons. *Conoids and Spheroids* comprises forty propositions on volumes of paraboloids, prolate spheroids (football-shaped), oblate spheroids (Earth-shaped), and other quadric surfaces. He also wrote on *Semi-regular Polyhedrons*, now unfortunately lost. In Euclid's *Elements* it is proven that there are five regular polyhedra, each made up of identical regular polygons. Archimedes extended this by discovering eleven additional solids made up of regular but dissimilar polygons. (Now it is known that there are a total of thirteen such Archimedean solids.)

The *Sandreckoner* is a treatise on how to write arbitrarily large numbers. In it he classified large numbers by order. First order were numbers from 1 to 10^8. Second order were numbers from

10^8 to 10^{16}. Third-order numbers are those from 10^{16} to 10^{24}, and so on. Our modern numbering system breaks up numbers by groups of 10^3 with thousands, millions, billions, trillions, etc. In Japan, numbers are broken up by groups of 10^4. Archimedes's system was based on powers of 10^8. He determined that 10^{51} grains of sand would be enough to fill "the universe"—the space within the "orbit" of the sun about the Earth.

And there are many more essays—*The Book of Lemmas* and *The Method* are still extant, but *Geometrical Methods, Parallel Lines, Triangles, Properties of Right Angled Triangles, Data, The Heptagon Inscribed in a Circle,* and *Systems of Circles Touching One Another* are all lost.

An intriguing problem due to Archimedes is the "stomachion" or "loculus of Archimedes." It literally involves *fractions* of a square! The problem consists of a fourteen-piece jigsaw-type puzzle to which various shaped triangles, quadrilaterals, and pentagons have to be fitted together to form a perfect square (Figure 13.2). The problem is not just to find one solution (hard enough), but to actually enumerate all such possible solutions. This problem was

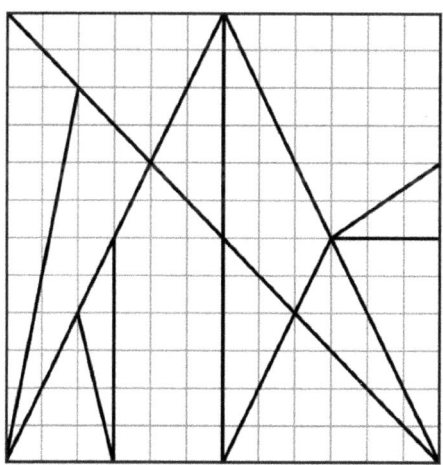

Figure 13.2 Stomachion—a fourteen-piece jigsaw puzzle attributed to Archimedes.
Reproduced from Yves Lemarcheix (2009) via Wikimedia Commons.

miraculously rediscovered in the early 1900s on a palimpsest, a piece of parchment that had been written over during the Middle Ages. It wasn't until 2003 that that Bill Cutler found all 536 possible distinct arrangements of the pieces disallowing reflections and rotations.

Archimedes wrote three significant treatises on plane geometry. The first, *On Spirals*, includes twenty-eight propositions using the "spiral of Archimedes" to both trisect a given angle and to square a circle. Along with duplicating the cube, Greek mathematicians sought the simplest possible methods to solve each of these three famous geometric problems of antiquity. In the nineteenth century it was shown that none of the problems can be solved using only the Euclidean tools of straightedge and compass. The second treatise, *The Quadrature of the Parabola*, contains twenty-four propositions highlighted by constructing a triangle having 3/4 the area of a parabolic segment. This was the first quadrature of a conic section and demonstrated a highly facile and sophisticated utilization of his method of exhaustion. His insights and methods inspired mathematicians centuries later in their development of integral calculus and infinite series. Briefly, we now turn our attention to the third work on plane geometry, *The Measurement of the Circle*, which has but three propositions. Proposition 3 states:

> The ratio of the circumference of any circle to its diameter is less than $3\frac{1}{7}$ but greater than $3\frac{10}{71}$.

Here we have a significant advancement in humanity's estimate of the value of pi. In 1 Kings 7.23 of the Old Testament, it is stated that, "And he made a molten sea, ten cubits from the one brim to the other: it was round all about, and his height was five cubits; and a line of thirty cubits did compass it round about." We can infer that the ancient Hebrew scholars used a value of pi of 3 since that is the ratio of the circumference of a circle to its diameter. In a contemporaneous Egyptian work, the area of a circle of diameter d is taken to be the same as that of a square with edges of length $(8/9)d$. Since the ratio of the area of a circle to the square of its radius is pi and the diameter is twice the radius, we calculate that the area of the square would be

$$\left(\frac{8}{9}d\right)^2 = \frac{64}{81}d^2 = \frac{256}{81}r^2,$$

where d is the diameter of the circle and r its radius. Since the area of a circle is given by $A = \pi r^2$, the Egyptian value of pi was $\frac{256}{81}$ = 3.16049. . . (a slight overestimate advantageous to the tax collector who may have used this estimate in land measurement).

Archimedes' method was to take a unit circle (having radius 1 and circumference 2π) and make careful estimates for the perimeters of regular inscribed and circumscribed polygons of ever increasing number of sides to gradually "trap" pi. He began with a circumscribed hexagon, then doubled the number of sides repeatedly until he reached a circumscribed ninety-six-gon. The perimeter of the resulting polygon was a very close overestimate for the circumference of the enclosed circle. Then he carried out similar calculations with inscribed polygons starting with a hexagon and proceeding step by step to an inscribed ninety-six-gon to get a good underestimate for the circumference of the circle. In Figure 13.3 we show the initial diagram for one side of the circumscribed hexagon.

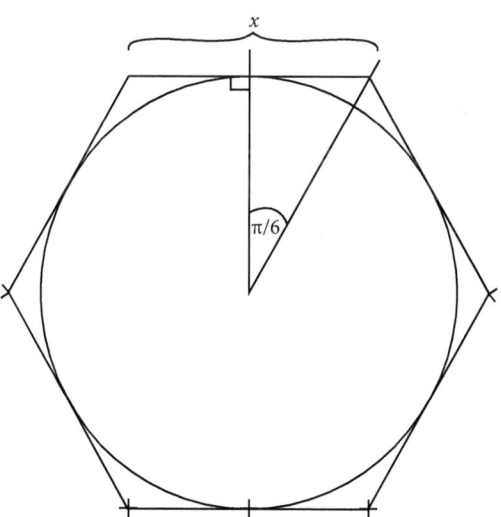

Figure 13.3 One step in constructing a circumscribed hexagon.

In the figure, the radius of the circle is 1, the central angle shown is $\pi/6$ and we let x be the length of one side of the hexagon. Using basic trigonometry,

$$\tan \frac{\pi}{6} = \frac{x/2}{1},$$

which implies that $x = \frac{2}{\sqrt{3}}$ since

$$\tan \frac{\pi}{6} = \frac{1}{\sqrt{3}}.$$

At this stage of the analysis, Archimedes already needed upper and lower bounds for $\sqrt{3}$. Somehow, he chose

$$\frac{265}{153} < \sqrt{3} < \frac{1351}{780}.$$

By the way, $\frac{265}{153} = 1.732026\ldots$, $\sqrt{3} = 1.732050\ldots$, and $\frac{1351}{780} = 1.732051\ldots$. (I remember the initial part of the decimal expansion of $\sqrt{3}$ by simply remembering the birth year of George Washington—1732. I guess this wouldn't help Archimedes much.)

Notice that Archimedes wanted to choose fractions near to $\sqrt{3}$. To do so, the quantity $a^2 - 3b^2$ must be small. In this case, $265^2 - 3(153)^2 = -2$ and $1351^2 - 3(780)^2 = 1$. He certainly chose wisely. But how did he come up with these particular estimates? Unfortunately, he didn't say. Here is a very plausible explanation.

Recall formula (12.4) from Chapter 12. Let $x = A_0$ be any irrational number. Set $a_0 = [A_0]$, the greatest integer less than or equal to A_0. Let

$$A_1 = \frac{1}{A_0 - a_0}$$

and $a_1 = [A_1]$, and

$$A_n = \frac{1}{A_{n-1} - a_{n-1}} \quad \text{and} \quad a_n = [A_n] \text{ for all } n \geq 1.$$

Also $r = \sqrt{3} = A_0$ and $a_0 = \left[\sqrt{3}\right] = 1$, and

$$A_1 = \frac{1}{A_0 - a_0} = \frac{1}{\sqrt{3} - 1} = \frac{\sqrt{3} + 1}{2}.$$

So $a_1 = [A_1] = 1$ since

$$1 \le \frac{\sqrt{3} + 1}{2} < 2,$$

then

$$A_2 = \frac{1}{A_1 - a_1} = \frac{1}{\frac{\sqrt{3}+1}{2} - 1}$$

$$= \frac{2}{\sqrt{3} - 1} = \frac{2\left(\sqrt{3} + 1\right)}{\left(\sqrt{3} - 1\right)\left(\sqrt{3} + 1\right)} = \frac{2 + 2\sqrt{3}}{2} = 1 + \sqrt{3}.$$

Hence, $a_2 = [A_2] = 2$ since $2 \le 1 + \sqrt{3} < 3$, and

$$A_3 = \frac{1}{A_2 - a_2} = \frac{1}{\sqrt{3} - 1} = A_1.$$

From this point onwards, the partial quotients follow an alternating repeating pattern.

Hence, $a_1 = a_3 = a_5 = \ldots = 1$ and $a_2 = a_4 = a_6 = \ldots = 2$. Therefore, $\sqrt{3} = [1; 1, 2, \overline{1, 2}]$. Next, we work out the convergents to $\sqrt{3}$ via our usual tabulation method:

i	-2	-1	0	1	2	3	4	5	6	7	8	9	10	11
a_i			1	1	2	1	2	1	2	1	2	1	2	1
p_i	0	1	1	2	5	7	19	26	71	97	265	362	989	1351
q_i	1	0	1	1	3	4	11	15	41	56	153	209	571	780

Please notice that the eighth convergent for $\sqrt{3}$ is $\frac{265}{153}$ and the eleventh is $\frac{1351}{780}$. Eureka!

Exercises

1. The Chinese mathematician Liu Hui (third century CE) derived the value of 3.14159 for π by considering a regular polygon with 3,072 sides. Factor 3,072 to see how it might be related to Archimedes' estimate with ninety-six-sided polygons.

2. The French mathematician François Viète (1540–1603) derived an approximation to π accurate to ten decimal places by considering a regular polygon with 393,216 sides. See if you can factor this number. This was pretty much the limit of the Archimedean method of determining the value of π. Powerful analytic techniques replaced geometric methods from this point on.

3. Calculate the decimal fraction for the π estimate $\frac{3927}{1250}$ used by the great Indian algebraist Bhaskara (1117–85). Bhaskara solved many examples of Pell's equation $x^2 = ny^2 + 1$, including the smallest solution when $n = 67$, namely $x = 48,842$ and $y = 5,967$.

Chapter 14

Crash Course on Infinite Series

> I asked my wife, "On a scale of 1 to 10, how do you rate me as a lover?" She said, "You know I'm no good at fractions."
> **Rodney Dangerfield (1921–2004)**

Most of the material until now has involved finite sums of fractions. But infinite sums are full of delight as well. In fact, we've already discussed some interesting examples back in Chapters 2 and 3 when discussing geometric series and Fibonacci numbers. As expected, the mathematical concepts get more involved. In this chapter, we cover a stripped-down version of some of the material that would be covered in a standard calculus sequence relating to infinite series. The material will pay further dividends in Chapter 15.

Any infinite sum of numbers is known as an infinite series (or simply *series*). An essential aspect of any series is whether it converges (to a finite sum) or diverges. Furthermore, and of greater difficulty, is to determine the precise sum of a convergent series. A general series is of the form $a_1 + a_2 + a_3 + a_4 + \ldots$, where the sum goes on forever. Typically, this is abbreviated using sigma notation:

$$\sum\nolimits_{i=1}^{\infty} a_i = a_1 + a_2 + a_3 + a_4 + \ldots.$$

We can get a good feel for any series by computing some of its initial *partial sums*.

The nth partial sum s_n is simply the sum of the first n terms of the series

$$s_n = \sum\nolimits_{i=0}^{n} a_i = a_1 + a_2 + \ldots + a_n.$$

For example, let's look at some simple series.

Example 1: $\sum_{i=1}^{\infty} i^2$. The series begins $1 + 4 + 9 + 16 + 25 + \ldots$ The first few partial sums are $s_1 = 1$, $s_2 = 5$, $s_3 = 14$, $s_4 = 30$, and so on. Each partial sum adds the next square on to the previous partial sum. There is a well-known formula for any of these finite sums. But in any event, it's clear that the partial sums will simply get larger and larger without bound. Hence, the series of squares diverges.

Example 2: $\sum_{i=1}^{\infty} \frac{1}{2^i}$. The series begins $\frac{1}{2} + \frac{1}{4} + \frac{1}{8} + \frac{1}{16} + \ldots$ with partial sums $s_1 = \frac{1}{2}$, $s_2 = \frac{3}{4}$, $s_3 = \frac{7}{8}$, $s_4 = \frac{15}{16}$, and so on. In this case, we would be correct to guess that the sums are getting closer and closer to 1. In fact, this is a geometric series which we've covered fully in Chapter 2. The initial term is a $= \frac{1}{2}$ and the ratio of successive terms is $r = \frac{1}{2}$. By our previous result, it converges to $\frac{a}{1-r} = 1$. Alternatively, notice that the partial sums $s_n = 1 - \frac{1}{2^n}$ and that as 2^n grows arbitrarily large, the fraction $\frac{1}{2^n}$ gets arbitrarily closer and closer to 0. The sum of the series is the limit as n goes to infinity of the partial sums. We write the sum as $S = \lim_{n \to \infty} s_n = 1$.

Example 3: $\sum_{i=1}^{\infty} \frac{1}{i}$. This series begins $1 + \frac{1}{2} + \frac{1}{3} + \frac{1}{4} + \ldots$ with partial sums $s_1 = 1$, $s_2 = \frac{3}{2}$, $s_3 = \frac{11}{6}$, $s_4 = \frac{25}{12}$, etc. The sums are increasing slowly. Eleven terms are required for the partial sum to exceed 3 and thirty-one terms are needed to surpass 4. Even the sum of the first one million terms barely exceeds 14. It's not at all clear if the series is converging toward some ultimate sum or whether the series diverges, exceeding any finite limit. We need to analyze further.

Example 4: $\sum_{i=0}^{\infty} (-1)^n = 1 - 1 + 1 - 1 + \ldots$ The partial sums alternate $s_1 = 1$, $s_2 = 0$, $s_3 = 1$, $s_4 = 0$, and so on ping ponging between 0 and 1. Without a careful definition of convergence, it might not be so clear how to classify this series. In fact, historically there has been some confusion determining its convergence status.

Example 5: $\sum_{i=1}^{\infty} \frac{1}{i^2}$. The series begins

$$1 + \frac{1}{4} + \frac{1}{9} + \frac{1}{16} + \frac{1}{25} + \ldots, \tag{14.1}$$

which is the sum of the reciprocals of the squares. The first few partial sums are $s_1 = 1, s_2 = \frac{5}{4} = 1.25, s_3 = \frac{49}{36} = 1.3611\overline{1}, s_4 = \frac{205}{144} = 1.42361\overline{1}, s_5 = \frac{5126}{3600} = 1.42388\overline{8}$. Again, the sums are increasing ever so gradually. We will have more to say about this series shortly.

Since this is not a calculus textbook, I have avoided much of the technical detail and level of precision that would be required there. But at this point, we need to clarify convergence and divergence a bit further. A series converges if its sequence of partial sums has a finite limit. If it does, then this limit S is the *sum* of the series. The idea is that the partial sums get ever closer to S as we add more terms of a convergent series. More precisely, a series *converges to S* if given any positive number ε (epsilon), no matter how small, there is a positive integer N (perhaps dependent on ε) such that if $n \geq N$, then the difference in absolute value between S and s_n is strictly less than ε.

You can think of this as a two-person game. Suppose we are given a convergent series $\sum_{n=1}^{\infty} a_n$ with sum S. If you challenge me with a small number ε, I can then respond with a large integer N for which $|S - s_n| < \varepsilon$ for all $n \geq N$. If you give me another number ε, then I respond with a new N, and so on. Since the series converges, I can always win this game. But if the series $\sum_{n=1}^{\infty} a_n$ doesn't converge (i.e., diverges), then you can find some value of ε for which there is no N satisfying $|S - s_n| < \varepsilon$ for all $n \geq N$. I cannot come up with an appropriate N and you win the game.

Let's look back at our examples. As we already noted, the series in Example 1 diverges since the sequence of partial sums shoots right by any number S chosen. You could say that the series diverges to infinity.

Example 2 was a convergent geometric series with sum $S = 1$. If you challenge me with a small number $\varepsilon > 0$, I can find an N large enough for which $1 - s_n < \varepsilon$ for all $n \geq N$.

In Example 3, the series $\sum_{i=1}^{\infty} \frac{1}{i}$ is known as the *harmonic series*. Even though its partial sums grow ever more slowly, it was identified as being a divergent series via a clever argument by the Parisian scholar, and later Bishop of Lisieux, Nicole Oresme (1323–82) in his treatise *Tractacus de Figuratione Potentiarum et Mensurarum*. Oresme's idea was the following. After the first term, group the terms by powers of 2. Specifically,

$$\sum_{i=1}^{\infty} \frac{1}{i} = 1 + \frac{1}{2} + \left(\frac{1}{3} + \frac{1}{4} \right) + \left(\frac{1}{5} + \frac{1}{6} + \frac{1}{7} + \frac{1}{8} \right) + \left(\frac{1}{9} + \ldots + \frac{1}{16} \right)$$
$$+ \left(\frac{1}{17} + \ldots + \frac{1}{32} \right) + \ldots .$$

But

$$\frac{1}{3} + \frac{1}{4} > \frac{1}{4} + \frac{1}{4} = \frac{2}{4} = \frac{1}{2},$$

$$\frac{1}{5} + \frac{1}{6} + \frac{1}{7} + \frac{1}{8} > \frac{1}{8} + \frac{1}{8} + \frac{1}{8} + \frac{1}{8} = \frac{4}{8} = \frac{1}{2},$$

$$\frac{1}{9} + \ldots + \frac{1}{16} > \frac{1}{16} + \ldots + \frac{1}{16} = \frac{8}{16} = \frac{1}{2},$$

$$\frac{1}{17} + \ldots + \frac{1}{32} > \frac{1}{32} + \ldots + \frac{1}{32} = \frac{16}{32} = \frac{1}{2}, \text{ etc.}$$

Hence, the harmonic series

$$\sum_{i=1}^{\infty} \frac{1}{i} > 1 + \frac{1}{2} + \frac{1}{2} + \frac{1}{2} + \frac{1}{2} + \frac{1}{2} + \ldots$$

and the latter series clearly diverges.

We have compared the harmonic series with a series that, although term by term smaller, still diverges. Utilizing what is called the *comparison* test, the harmonic series must also diverge.

Another way to see why the harmonic series diverges is to consider the opposite of what we wish to prove and show that the false hypothesis leads to a contradiction. Suppose the harmonic series converges to S:

$$S = 1 + \frac{1}{2} + \frac{1}{3} + \frac{1}{4} + \frac{1}{5} + \frac{1}{6} + \frac{1}{7} + \dots.$$

Multiply both sides by ½:

$$\frac{S}{2} = \frac{1}{2} + \frac{1}{4} + \frac{1}{6} + \frac{1}{8} + \dots.$$

Subtracting the second equation from the first obtains

$$S - \frac{S}{2} = \frac{S}{2} = 1 + \frac{1}{3} + \frac{1}{5} + \frac{1}{7} + \dots.$$

In this case, the sum of the reciprocals of the odd natural numbers would equal the sum of the reciprocals of the even natural numbers. However, $1 > \frac{1}{2}, \frac{1}{3} > \frac{1}{4}, \frac{1}{5} > \frac{1}{6}, \frac{1}{7} > \frac{1}{8}$, etc., which implies that the sum of reciprocals of the odd numbers must strictly exceed that for the even numbers. This is a contradiction. Hence, our hypothesis that the series converges to some number S is incorrect, and so the harmonic series must diverge.

While we're at it, here's a nice popular mathematical puzzle. Show that none of the partial sums of the harmonic series (beyond the first) is an integer. To solve this puzzle, note that a rational number $\frac{m}{n}$ reduces to being an integer if and only if the denominator n divides the numerator m. Suppose there is a partial sum $s_r = 1 + \frac{1}{2} + \dots \frac{1}{r}$ equalling an integer. Let 2^k be the largest power of 2 for which $2^k \le r$ $< 2^{k+1}$ with $k \ge 1$. Let L be the least common multiple of the integers $1, 2, \dots, r$. Then 2^k divides L, but 2^{k+1} does not divide L. (We say that 2^k *exactly* divides L.) Rewriting s_r with this common divisor, we obtain

$$s_r = \frac{L + L/2 + L/3 + L + L/r}{L}.$$

Each term of the numerator is even save for a single term, namely the 2^k-th term. Hence the sum in the numerator is odd. But the denominator is even. Therefore, the denominator does not divide the numerator, contradicting the assumption that s_r is an integer.

Example 4 is an example of an *alternating* series since we alternate between adding and subtracting various quantities. One could argue erroneously that it converges to either 0 or 1 or perhaps converges to both 0 and 1. At one time, it was argued that we should say it converges to their average value, ½. However, the series does not satisfy our definition of convergence, and hence we must conclude that this series diverges. Going back to the convergence game alluded to previously, for any choice of $\varepsilon < \frac{1}{2}$, no matter what choice of S we posit and any N as large as we wish, there will always be partial sums s_n with $n > N$ for which

$$|S - s_n| > \varepsilon.$$

The divergence of the series in Examples 1 and 4 can also be demonstrated by use of the following key result.

The nth Term Divergence Test: Given an infinite series $\sum_{n=1}^{\infty} a_n$, if $\lim_{n \to \infty} a_n \neq 0$ or doesn't exist, then the series $\sum_{n=1}^{\infty} a_n$ diverges.

Quick proof: We establish the contrapositive of the nth term divergence test which is equivalent to it. Suppose that $\sum_{n=1}^{\infty} a_n$ converges. Let S be its sum. Note that $a_n = s_n - s_{n-1}$ where s_n is the nth partial sum of the series. Hence

$$\lim_{n \to \infty} a_n = \lim_{n \to \infty} (s_n - s_{n-1}) = \lim_{n \to \infty} s_n - \lim_{n \to \infty} s_{n-1} = 0.$$

In the proof above, we used the fact that if the limits exist, the limit of a sum is the sum of the limits.

In Example 1, $\lim_{n \to \infty} n^2 = \infty$ (actually doesn't exist) and in Example 4, $\lim_{n \to \infty} (-1)^n$ also doesn't exist. Hence both series diverge.

Please note that the converse of the nth term divergence test does not hold generally. In both Examples 2 and 3, the terms of the series approached 0. Yet in the first case the series converged, while in the second case it did not. The nth term approaching 0 is a necessary condition for convergence, not a sufficient one.

The fifth example, equation (14.1), is the thorniest of our series thus far considered. It was first studied by the Swiss mathematician Jakob Bernoulli (1654–1705), who determined that it converges. His method was to compare it with a known convergent series that was term by term larger. The comparison series used was

$$1 + \sum_{n=1}^{\infty} \frac{1}{n(n+1)} = 1 + \frac{1}{2} + \frac{1}{6} + \frac{1}{12} + \frac{1}{20} + \dots, \qquad (14.2)$$

the denominators being what the ancient Greeks would call *oblong* numbers (rectangles that were nearly square). Rewrite

$$\frac{1}{n(n+1)} \quad as \quad \frac{1}{n} - \frac{1}{n+1}$$

to obtain

$$1 + \sum_{n=1}^{\infty} \frac{1}{n(n+1)} = \sum_{n=1}^{\infty} \frac{1}{n} - \frac{1}{(n+1)} = 1 + \left(\frac{1}{1} - \frac{1}{2}\right)$$
$$+ \left(\frac{1}{2} - \frac{1}{3}\right) + \left(\frac{1}{3} - \frac{1}{4}\right) + \left(\frac{1}{4} - \frac{1}{5}\right) + \dots.$$

This is known as a telescoping or collapsing sum since consecutive terms cancel. In this case, the nth partial sum

$$s_n = 1 + \left(1 - \frac{1}{n+1}\right).$$

Since $\lim_{n \to \infty} s_n = 2$, the series (14.2) converges to 2. Thus, by the comparison test, the series (14.1) must also converge. But what is its sum S?

Several mathematicians studied it. John Wallis (1616–1703), the Savilian Professor of Geometry at Oxford University, calculated its sum to three decimal places, obtaining $S \approx 1.645$, though he was uncertain whether it was accurate even to that level. Christian

Figure 14.1 Leonhard Euler, painted by Jakob Emanuel Handmann (1753).
Reproduced from Kunstmuseum Basel via Wikimedia Commons.

Goldbach (1690–1764), a German mathematician working in Russia at the Academy at St. Petersburg, was able to verify that

$$1 + \frac{16,223}{25,200} = 1.64376984\ldots < S < 1 + \frac{30,197}{46,800} = 1.64523504\ldots.$$

Jakob Bernoulli's younger brother, Johann (1667–1748), with whom he had an intense rivalry also worked on the problem without full success. The problem of determining the exact sum of the series of square reciprocals became known as the Basel Problem since Basel was the city in which it originated. Eventually, the problem was solved by a student of Johann Bernoulli's, an individual destined to become the greatest mathematician of the eighteenth century.

The Swiss mathematician Leonhard Euler (1707–83) (Figure 14.1) was able to demonstrate conclusively that in fact

$S = \frac{\pi^2}{6} = 1.644934...$, a stunning achievement for which he immediately gained much fame throughout the scientific community. Euler then went on to study the sums of reciprocals of higher powers beyond squares. Define the *zeta function* by

$$\zeta(s) = \sum_{n=1}^{\infty} \frac{1}{n^s} \text{ for } s > 1,$$

Euler was able to evaluate $\zeta(s)$ for all positive even integers $s = 2k$. The formulas are a bit complicated. But each one is an irrational number of the form of a specific rational number times π^{2k}. The situation for odd powers is still far from fully understood. It wasn't until as recently as 1973 that the French mathematician Roger Apéry (1916–94) proved that $\zeta(3) = 1.2020569...$ is, in fact, irrational. A deep study of the zeta function was undertaken by the great German mathematician G. B. Riemann (1826–66) in 1859 where the argument s was extended for complex values of s. Now we refer to the complex-valued function more generally as the Riemann zeta function.

Integral calculus can be used to show that $\zeta(s)$ converges for $s > 1$. This is known as the *p-test*: The series $\sum_{n=1}^{\infty} \frac{1}{n^p}$ converges for all $p > 1$ and diverges otherwise. The p-test implies that the series $\sum_{n=1}^{\infty} \frac{1}{n^2}$, $\sum_{n=1}^{\infty} \frac{1}{n^{\sqrt{2}}}$, and $\sum_{n=1}^{\infty} \frac{1}{n^{100}}$ all converge while the series $\sum_{n=1}^{\infty} \frac{1}{n}$ and $\sum_{n=1}^{\infty} \frac{1}{\sqrt{n}}$ diverge. Here is a nifty conundrum to consider which I call the *box paradox*.

Box Paradox: Suppose you wish to paint the interior of an infinite number of cubical boxes of sides $\frac{1}{\sqrt{1}}, \frac{1}{\sqrt{2}}, \frac{1}{\sqrt{3}}, \frac{1}{\sqrt{4}}, ..., \frac{1}{\sqrt{n}}, ...,$ respectively. Each box has six square sides. Before constructing the boxes, you lay out all the sides on a flat surface in order to paint one side of each square piece. The total area painted is then six times the sum of the areas of one square piece of each size. Hence, the area painted is

$$6 \sum_{n=1}^{\infty} \left(\frac{1}{\sqrt{n}} \right)^2 = 6 \sum_{n=1}^{\infty} \frac{1}{n}.$$

But the harmonic series diverges, and hence we cannot possibly paint the sides of all the boxes. Perhaps this doesn't seem especially surprising. There are an infinite number of boxes after all.

But then we hit upon another idea. Construct all the boxes and then fill each one completely with paint. Then drain out any residue, just keeping the paint that has attached to the interior sides. The total amount of volume of paint required is

$$\sum_{n=1}^{\infty} \left(\frac{1}{\sqrt{n}}\right)^3 = \sum_{n=1}^{\infty} \frac{1}{n^{3/2}}.$$

By the p-test, since $3/2 > 1$, the last series converges and so only a finite amount of paint is required. How's that possible?!

Alternating series form a special but important and common class of infinite series. These are series where the terms a_i alternate between being positive and being negative. Compared to a general infinite series, determining the convergence status of an alternating series is much simpler. Recall that for a general series, by the divergence test, the nth term must approach 0. However, that was a necessary but not sufficient condition for convergence. For alternating series, we have an elegant theorem which we state without proof.

Alternating Series Test: If the alternating series $\sum_{n=1}^{\infty} (-1)^{n-1} b_n = b_1 - b_2 + b_3 - b_4 + \ldots$, where all $b_n > 0$ for all n satisfies

(a) $b_{n+1} \leq b_n$ for all n and
(b) $\lim_{n\to\infty} b_n = 0$,

then the series converges.

For example, the alternating harmonic series

$$1 - \frac{1}{2} + \frac{1}{3} - \frac{1}{4} + \ldots$$

must converge by the Alternating Series Test. Similarly, so does the series

$$1 - \frac{1}{3} + \frac{1}{5} - \frac{1}{7} + \ldots.$$

Shortly, we will determine the actual sum of these two series.

So far, we have studied series consisting of an infinite sum of numbers such as the harmonic series and the sum of reciprocals of the squares. But the notion of a series can be extended to adding an infinite sum of monomials as a function of x. Here are two representative examples:

$$\sum_{n=0}^{\infty} x^n = 1 + x + x^2 + x^3 + x^4 + \ldots \tag{14.3}$$

$$\sum_{n=0}^{\infty} (-1)^{n+1} \frac{x^{2n+1}}{2n + 1} = 1 - \frac{x^3}{3} + \frac{x^5}{5} - \frac{x^7}{7} + \ldots. \tag{14.4}$$

These *power series* extend polynomials from the finite realm into the infinite. We have already seen an example, formula (3.1), the generating function for the Fibonacci numbers. Each substitution of a real number for x results in a different particular infinite series. Some choices for x lead to convergent series; other choices may lead to divergent series. For example, "plugging in" $x = \frac{1}{2}$ in series (14.3) gives the convergent geometric series

$$1 + \frac{1}{2} + \frac{1}{4} + \frac{1}{8} + \ldots,$$

while $x = 3$ leads to the divergent series

$$1 + 3 + 9 + 27 + 81 + \ldots.$$

A generic power series is of the form $\sum_{n=0}^{\infty} c_n(x - a)^n$. For the power series in formula (14.3), $a = 0$ and $c_n = 1$ for all n. In formula (14.4), $a = 0$ and $c_{2k} = 0$ while

$$c_{2k+1} = \frac{(-1)^k}{2k + 1} \text{ for } k \geq 0.$$

Given a power series, it is of interest to determine the set of values of x for which it converges. In all cases, the set of convergent values is an interval centered at $x = a$. When $x = a$, the series trivializes to an infinite sums of zeros. The *interval of convergence* might just be the value $x = a$, some finite (open, closed, half-open) interval centered at a, or even the entire real line. The distance from a to the edge of the interval of convergence is known as the *radius of* convergence, denoted by the letter R. When the power series only converges at $x = a$, we say that $R = 0$. On the opposite extreme, $R = \infty$ when the series converges for all x. In between, R is some positive constant. In that case, what happens at the two boundaries of the interval of convergence depends on the particular power series. We'll save a full discussion of this to a good calculus course. However, please note that it is legitimate to add or multiply two convergent power series together just as we might for polynomials as long as we constrain ourselves to values of x within the intersection of both intervals of convergence. For those of you who know some calculus, another key result is that a power series can be either differentiated or integrated term by term within its interval of convergence. Let's state this more precisely.

Power Series Theorem: If the power series $\sum_{n=0}^{\infty} c_n(x - a)^n$ has radius of convergence $R > 0$, then the function defined by $f(x) = \sum_{n=0}^{\infty} c_n(x - a)^n$ is differentiable on the interval $(a - R, a + R)$ and

(a) $f'(x) = c_1 + 2c_2(x - a) + 3c_3(x - a)^2 + 4c_4(x - a)^3$
$\quad\quad + 5c_5(x - a)^4 + \ldots$

(b) $\int f(x)dx = c_0(x - a) + c_1\frac{(x-a)^2}{2} + c_2\frac{(x-a)^3}{3} + c_4\frac{(x-a)^4}{4} + \ldots + C,$

with the series in (a) and (b) both having radii of convergence R.

Here is an extended example that utilizes the Power Series Theorem. Recall that the geometric series $a + ar + ar^2 + ar^3 + \ldots$ converges for $|r| < 1$ with sum $\frac{a}{1-r}$. By letting $a = 1$ and $r = x$, this implies that the geometric power series $1 + x + x^2 + x^3 + \ldots$ converges to $\frac{1}{1-x}$ whenever $|x| < 1$. Hence, power series (14.3) has radius of convergence $R = 1$ and, in fact, its interval of convergence is $(-1, 1)$. If we replace x with

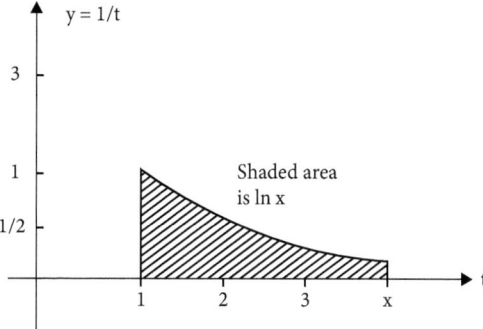

Figure 14.2 Graph of $y = 1/t$ with area $\ln x$ shaded.

$-x$, then we obtain

$$\frac{1}{1 + x} = 1 - x + x^2 - x^3 + \ldots \text{valid where } \left|- x\right| = |x| < 1. \quad (14.5)$$

Within that interval, we can integrate to obtain

$$\int \frac{1}{1 + x}\,dx = x - \frac{x^2}{2} + \frac{x^3}{3} - \frac{x^4}{4} + \ldots + C \text{ for } |x| < 1 \,|$$

where C is a constant of integration.

Although the Power Series Theorem only guarantees its convergence within the interval $(-1, 1)$, we know from the Alternating Series Test that we can extend the convergence to the boundary value $x = 1$. The integral is readily evaluated as

$$\int \frac{1}{1 + x}\,dx = \ln|1 + x| + C.,$$

where $\ln(x)$ is the natural logarithm function. When $x > 1$, $\ln(x)$ represents the area from $t = 1$ to $t = x$ below the graph of the function $y = 1/t$ and above the horizontal t-axis (Figure 14.2).

When $x = 0$, $\ln|1 + x| = \ln(1) = 0$ and hence $C = 0$ in this case. Letting $x = 1$ leads to a convergent alternating series called the alternating harmonic series. Therefore,

$$\ln 2 = 1 - \frac{1}{2} + \frac{1}{3} - \frac{1}{4} + \ldots = \sum_{n=1}^{\infty} (-1)^{n+1}\frac{1}{n} \approx 0.693147.\ldots$$

Returning to formula (14.5), substitute x^2 for x to obtain

$$\frac{1}{1+x^2} = 1 - x^2 + x^4 - x^6 + \dots \text{valid where } |x| < 1.$$

Integrating term by term obtains

$$\int \frac{1}{1+x^2} dx = x - \frac{x^3}{3} + \frac{x^5}{5} - \frac{x^7}{7} + \dots + C.$$

But

$$\int \frac{1}{1+x^2} dx = \tan^{-1}(x) + C,$$

the arctangent function (think of it as the angle whose tangent is x). Since $\tan^{-1}(0) = 0$, $C = 0$. Thus,

$$\tan^{-1}(x) = x - \frac{x^3}{3} + \frac{x^5}{5} - \frac{x^7}{7} + \dots \text{for } |x| < 1. \qquad (14.6)$$

What about when $x = 1$? In this case, the series on the right is an alternating series that satisfies the conditions prescribed in the Alternating Series Test. So the series also converges for $x = 1$. But the tangent function at $\frac{\pi}{4}$ is 1; hence, $\tan^{-1}(1) = \frac{\pi}{4}$. Substituting 1 in equation (14.6) obtains

$$\frac{\pi}{4} = 1 - \frac{1}{3} + \frac{1}{5} - \frac{1}{7} + \dots.$$

This is a very pretty result first discovered by the Scottish astronomer and mathematician James Gregory (1638–75) in 1671. It should be mentioned that Gottfried Wilhelm Leibniz (1646–1716) independently discovered the same result. In fact, it may have been known to the Indian astronomer and mathematician Kerala Nilikantha Somayaji in the fifteenth century!

The Ratio Test is an especially useful test for determining the radius of convergence of a series. Given an infinite series

$$\sum_{n=1}^{\infty} a_n, \text{ let } L = \lim_{n \to \infty} \left| \frac{a_{n+1}}{a_n} \right|.$$

The Ratio Test states that if $L < 1$, then the series converges and if $L > 1$, then the series diverges. (The case $L = 1$ is inconclusive.)

Here's a typical example: Find the interval of convergence of the series $\sum_{n=0}^{\infty} (-1)^n 2^n x^n$.

Solution: In this case,

$$L = \lim_{n \to \infty} \left| \frac{(-1)^{n+1} 2^{n+1} x^{n+1}}{(-1)^n 2^n x^n} \right| = \lim_{n \to \infty} |2x| = 2\,|x|\,.$$

By the Ratio Test, the series converges when $|x| < \frac{1}{2}$ and diverges when $|x| > \frac{1}{2}$. So its radius of convergence $R = \frac{1}{2}$. In other words, the series converges on the interval $\left(-\frac{1}{2}, \frac{1}{2}\right)$, and diverges when $x < -\frac{1}{2}$ or when $x > \frac{1}{2}$. The cases $x = -\frac{1}{2}$ and $x = \frac{1}{2}$ must be dealt with separately.

When $x = -\frac{1}{2}$, the series becomes

$$\sum_{n=0}^{\infty} (-1)^n 2^n \left(-\frac{1}{2}\right)^n = \sum_{n=0}^{\infty} 1 = 1 + 1 + 1 + \dots,$$

which clearly diverges.

When $x = \frac{1}{2}$, the series becomes

$$\sum_{n=0}^{\infty} (-1)^n 2^n \left(\frac{1}{2}\right)^n = \sum_{n=0}^{\infty} (-1)^n = 1 - 1 + 1 - 1 + \dots,$$

which also diverges by the nth term divergence test. Hence, the interval of convergence of the series $\sum_{n=0}^{\infty} (-1)^n 2^n x^n$ is $\left(-\frac{1}{2}, \frac{1}{2}\right)$.

As a final exercise, let's combine several of our results to study a thinned-out version of the harmonic series. Consider the series of reciprocals of all positive integers not having the digit 7 in its decimal representation. This series begins as follows:

$$1 + \frac{1}{2} + \frac{1}{3} + \dots + \frac{1}{6} + \frac{1}{8} + \dots + \frac{1}{16} + \frac{1}{18} + \dots + \frac{1}{69} + \frac{1}{80} + \dots, \quad (14.7)$$

where all numbers containing any 7s are plucked out of the harmonic series. Call these numbers the 7-free numbers. An infinite number of terms still remain. Previously, we established that the harmonic series diverges. Perhaps surprisingly, series (14.7) actually converges!

Proposition *(A. J. Kempner, 1917): The series* $\sum_{n=1(7-free)}^{\infty} \frac{1}{n}$ *converges.*

Proof: Since all digits from 1 to 9 save for 7 itself is 7-free, there are 8 one-digit positive integers that are 7-free. There are $8 \cdot 9 = 72$ two-digit numbers that are 7-free. (The first digit can be any digit from 1 to 9 except 7 and the second digit can by any digit from 0 to 9 except 7.) Similarly, there are $8 \cdot 9 \cdot 9 = 648$ three-digit numbers that are 7-free, and in general, there are $8 \cdot 9^{n-1}$ 7-free n-digit numbers. Furthermore, if n is a one-digit number, then $1/n \le 1$. If n is a two-digit number, then $1/n \le 1/10$. In general, if n is a k-digit number, then $1/n \le 1/10^{k-1}$. Hence

$$\sum_{n=1(7-free)}^{\infty} \frac{1}{n} < (1 + \ldots + 1) + \left(\frac{1}{10} + \ldots + \frac{1}{10}\right)$$

$$+ \left(\frac{1}{100} + \ldots + \frac{1}{100}\right) + \ldots$$

$$= \sum_{k=1}^{\infty} 8 \cdot 9^{k-1} \left(\frac{1}{10}\right)^{k-1}$$

$$= 8 \sum_{k=1}^{\infty} \left(\frac{9}{10}\right)^{k-1}$$

The last expression is a convergent geometric series ($r = 9/10 < 1$) and hence the series $\sum_{n=1(7-fres)}^{\infty} \frac{1}{n}$ converges.

There is nothing more special about the number 7 than any other digit. The proposition is still true if we replace the sum with 4- or 5-free numbers and so on. Looking from an opposite perspective, if we start with the harmonic series and remove all numbers that do not contain a 7 (i.e., that are not 7-free), then that series must diverge since we are removing a finite sum from the divergent harmonic series. The sum will still be infinite. That is, the sum of reciprocals of all numbers that are not 7-free diverges. And similarly, the same is still true if we then remove all 1-free numbers, then 2-free numbers, etc., since each thinning out just takes away a finite sum.

So what terms are left to add up to infinity? Remaining are the reciprocals of all numbers that contain all ten decimal digits. The smallest such number is 1,023,456,789. All smaller numbers are missing at least one of the ten digits. We deduce this charming corollary to the previous proposition.

Corollary: Let D denote the set of all positive integers having decimal expansions that contain all ten decimal digits. Then $\sum_{n \in D} \frac{1}{n}$ diverges.

In the world of infinite series, we have traveled a significant distance among its varied and winding paths. The paths themselves crisscross and seem to extend indefinitely. For example, it can be shown that the sum of reciprocals of the primes diverges (Euler, 1737), and thus that the sum of the reciprocals of just the primes which contain all ten decimal digits diverges as well. This latter sum begins with the reciprocal of 10,123,457,689, then 10,123,465,789, then 10,123,465,897. Even with all the primes not containing all ten decimal digits removed, we still have a divergent series!

Current research continues with related questions. As recently as 2016, the Oxford analytic number theorist and Fields medalist James Maynard (b. 1987) proved that the number of 7-free primes is itself infinite, and the same is true of n-free primes for any other digit. Lots to explore!

Exercises

1. Use the Ratio Test to determine the radius of convergence for the following series:

    (a) $\sum_{n=1}^{\infty} \frac{x^n}{2}$

    (b) $\sum_{n=1}^{\infty} \frac{x^n}{2^n}$

    (c) $\sum_{n=1}^{\infty} 2^n x^n$

    (d) $\sum_{n=1}^{\infty} (-1)^n n^2 x^n$.

2. Bertrand's Postulate is the theorem that for any integer $n \geq 2$, there is a prime p with $n < p < 2n$. Utilize Bertrand's Postulate to prove that, except for 1, none of the partial sums of the harmonic series is an integer.

3. Sum the following series due to Nicole Oresme:

   (a) $\sum_{n=1}^{\infty} \frac{n}{2^n}$

   (b) $\sum_{n=1}^{\infty} \frac{3n}{4^n}$.

4. In this problem, we will establish a stunning result due to Jakob Bernoulli:

$$\sum_{n=1}^{\infty} \frac{n^2}{2^n} = \frac{1}{2} + \frac{4}{4} + \frac{9}{8} + \frac{16}{16} + \frac{25}{32} + \ldots = 6.$$

   (a) Verify that the power series

$$\sum_{n=1}^{\infty} \left(\frac{x}{2}\right)^n.$$

   converges for $|x| < 2$.

   (b) Use the geometric series test to show that

$$\sum_{n=1}^{\infty} \left(\frac{x}{2}\right)^n = \frac{x}{2 - x}$$

   for $|x| < 2$.

   (c) For $|x| < 2$, differentiate the series in (b) term by term to show that

$$\sum_{n=1}^{\infty} \frac{n x^{n-1}}{2^n} = \frac{2}{(2 - x)^2}.$$

   (d) Let $x = 1$ to confirm an answer to exercise 14.3(a).

   (e) Multiply the equation in part (c) by x and then differentiate term by term to obtain

$$\sum_{n=1}^{\infty} \frac{n^2 x^{n-1}}{2^n} = \frac{2x + 4}{(2 - x)^3}.$$

   (f) Let $x = 1$ to obtain the sought-after result.

5. (a) Multiply the equation in problem 14.4(e) to obtain

$$\sum_{n=1}^{\infty} \frac{n^2 x^n}{2^n} = \frac{2x^2 + 4x}{(2 - x)^3}.$$

   (b) Differentiate the above term by term with respect to x.

   (c) Let $x = 1$ to determine

$$\sum_{n=1}^{\infty} \frac{n^3}{2^n}.$$

   (This is also due to Jakob Bernoulli.)

6. (a) Verify Catalan's Identity:

$$1 - \frac{1}{2} + \frac{1}{3} - \frac{1}{4} + \ldots + \frac{1}{2n - 1} + \frac{1}{2n} = \frac{1}{n + 1} + \frac{1}{n + 2}$$
$$+ \ldots + \frac{1}{2n} \text{ for all } n \geq 1.$$

   (b) Use Catalan's identity to show that if

$$\frac{m}{n} = 1 - \frac{1}{2} + \frac{1}{3} - \frac{1}{4} + \ldots + \frac{1}{59},$$

   then m is divisible by 89.

Chapter 15

Introduction to Taylor Series

The essence of mathematics is its freedom.

Georg Cantor (1845–1918)

A deeper understanding of infinite series requires a much greater background with the techniques and results of integral and differentiable calculus. It will now be necessary to assume that you have already had a full year course of calculus. We will simply offer a brief survey of the material necessary to get a handle on a special type of power series called a Taylor series. Our goal in this chapter is to gain greater insight into how Euler discovered the sum of the reciprocals of the squares. We begin with a bit of review.

The *derivative* of a function, when it exists, gives the rate of change of the given function. We won't define it here other than to say it involves a special limit. The derivative of a differentiable function gives the slope of the graph of the function at each appropriate value. The second derivative describes its concavity, and so on. Some of the main results of differential calculus deal with what happens when you combine functions by adding them, multiplying them, etc. The theorems have names like the sum rule, constant-multiple rule, product rule, quotient rule, power rule, and chain rule. The derivative of any polynomial or even rational function (quotient of polynomials) can be calculated by applying a combination of these results. In addition, it is important to know the derivatives of some transcendental (nonalgebraic) functions such as the six trigonometric functions and the logarithm and exponential functions. If appropriate, here are some examples that may help jog your memory:

$$\frac{d}{dx}\left(4x^3 + 5x^2 + 10\right) = 12x^2 + 10x$$

(utilizing the power rule, sum rule, constant-multiple rule),

$$\frac{d}{dx}\left(\frac{x+2}{x^2}\right) = \frac{x^2(1) - (x+2)2x}{x^4} = \frac{-x^2 - 4x}{x^4} = \frac{-x-4}{x^3}$$

(utilizing the quotient rule and power rule),

$$\frac{d}{dx}(4\sin x + \tan x) = 4\cos x + \sec^2 x$$

(utilizing basic trig derivative formulas), and

$$\frac{d}{dx}[\ln x]^3 = 3(\ln x)^2 \cdot \frac{1}{x} = \frac{3(\ln x)^2}{x}$$

$$\left(\text{utilizing the chain rule and the result that } \frac{d(\ln x)}{dx} = \frac{1}{x}\right).$$

The *definite integral* of a positive-valued continuous function represents an area bounded beneath the given function. The technical definition also involves a limiting process involving sums. The *indefinite integral* is simply the most general antiderivative or reverse of the derivative. There are some analogous basic formulas pertaining to integrals as there are for derivatives. In addition, there is much greater attention paid to *techniques* of integration. The Fundamental Theorem of Calculus clarifies precisely in what way the derivative and the definite integral are inverse operations. Its proof and development are credited to both Isaac Newton and Gottfried Wilhelm

Leibniz (1646–1716). To refresh your memory a bit further:

$$\int x^2 + 4x + 9 \, dx = \frac{1}{3}x^3 + 2x^2 + 9x + C$$

(utilizing the power and sum rules for integration),

$$\int \cos x \, dx = \sin x + C$$

(since $d(\sin x)/dx = \cos x$),

$$\int \sin x \, dx = -\cos x + C$$

(since $d(\cos x)/dx = -\sin x$),

$$\int e^{3x} dx = \frac{1}{3}e^{3x} + C$$

(since $d(e^{3x})/dx = 3e^{3x}$), and

$$\int_1^2 2x \, dx = x^2\big|_1^2 = 2^2 - 1^2 = 3$$

(utilizing the power rule for integration).

Given a function f having derivatives of all orders, we define its *first Taylor polynomial about x = a* to be the linear function $T_1(x) = f(a) + f'(a) \cdot (x - a)$. The first Taylor polynomial has the same function value and the same slope as $f(x)$ at $x = a$. The *second Taylor polynomial about x = a* is the quadratic function

$$T_2(x) = f(a) + f'(a) \cdot (x - a) + \frac{f''(x)}{2!}(x - a)^2.$$

Recall that $n!$ (*n* factorial) is the product $n! = 1 \cdot 2 \cdot \ldots \cdot n$. The second Taylor polynomial about $x = a$ has the same function value, the same slope, and the same concavity as does $f(x)$ at $x = a$.

In general, the *nth Taylor polynomial about x = a* is the nth degree polynomial

$$T_n(x) == f(a) + f'(a) \cdot (x - a) + \frac{f''(x)}{2!}(x - a)^2 + \ldots + \frac{f^{(n)}(x)}{n!}(x - a)^n.$$

Analogously, the nth Taylor polynomial about $x = a$ has the same function value and the same derivatives of all orders up to the nth order as does the function f itself. These polynomials are named after the English mathematician Brook Taylor (1685–1731), who described these polynomials and their extensions to infinite series in a book *Methodus Incrementorum Directa et Inversa* in 1715.

The importance of Taylor polynomials is that they form a sequence of better and better approximations to the function $f(x)$ for values of x near $x = a$. I tell my students that the situation is somewhat analogous to the work of a detective. Suppose a bank is robbed by a lone robber at a particular time who runs away on foot. The detective arrives at the bank shortly afterwards knowing where and when the robbery took place. The more information she can ascertain about which direction the robber ran, how fast, which street he might have turned onto next, and the like, the better chance she has of the robber being found and apprehended quickly.

An example might be instructive. Let $f(x) = \cos x$ and let $a = 0$. We are interested in the behavior of $f(x)$ near $x = 0$. The value $\cos(\pi/10)$ can be approximated by knowing information about the cosine function at $x = 0$. In this case, $f'(x) = -\sin x$ and $f''(x) = -\cos x$. Hence, $f(0) = 1, f'(0) = 0$, and $f''(0) = -1$. Thus, the second Taylor polynomial for $f(x) = \cos x$ about $x = 0$ is

$$T_2(x) = 1 - \frac{1}{2}x^2.$$

The function $T_2(x)$ has the same function value, slope, and concavity when $x = 0$ as does the cosine function. In this case,

$$T_2(\pi/10) = 1 - \frac{1}{2}\left(\frac{\pi}{10}\right)^2 \approx 0.95065,$$

while $\cos(\pi/10) \approx 0.95106$ rounded to five decimal places. See Figure 15.1 for a graphical representation.

For functions that are infinitely differentiable, the Taylor polynomials can be extended to the *Taylor series for f about x = a*

$$T(x) = \sum_{n=0}^{\infty} \frac{f^n(x)}{n!}(x-a)^n = f(a) + f'(a) \cdot (x-a) + \frac{f''(x)}{2!}(x-a)^2$$

$$+ \ldots + \frac{f^{(n)}(x)}{n!}(x-a)^n + \ldots .$$

The key issues for a Taylor series are (1) for what values of x does the series converge, and (2) where does the Taylor series $T(x)$ actually equal $f(x)$. Question (1) can often be answered via a convergence test such as the alternating series test (which we have discussed) or a more general convergence test such as the ratio test (which we also covered briefly). Question 2 is a bit thornier. However, the main idea is that the Taylor series actually equals the original function $f(x)$ if $f(x)$ equals the limit of the Taylor polynomials that approximate $f(x)$. Specifically, let $R_n(x) = f(x) - T_n(x)$ be the nth *remainder* of the Taylor series where $T_n(x)$ is the nth Taylor polynomial for f about $x = a$. If $\lim_{n\to\infty} R_n(x) = 0$ for $|x - a| < r$, then f is equal to its Taylor series on the interval $|x - a| < r$.

Luckily, in this chapter we will only be interested in the function $f(x) = \sin x$ which can be shown to be equal to its Taylor series for all x.

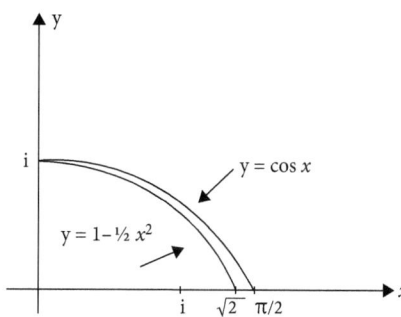

Figure 15.1 Second Taylor approximation to $y = \cos x$ about $x = 0$.

We now turn our attention to Euler's discovery of the exact value of

$$\zeta(2) = \sum_{n=1}^{\infty} \frac{1}{n^2}.$$

Euler was fully aware of all the algebraic, trigonometric, and calculus results that preceded him. In particular, he knew a theorem due to Newton: If a polynomial has constant term 1, then the sum of the reciprocals of its roots is the negative of the coefficient of the linear term. We begin by establishing Newton's result: Let

$$P(x) = c_n x^n + c_{n-1} x^{n-1} + \ldots + c_1 x + 1 \qquad (15.1)$$

be a polynomial with constant term 1. By the fundamental theorem of algebra (which was widely accepted but not quite rigorously proved at the time), $P(x)$ can be completely factored into n linear factors

$$P(x) = (a_1 x - r_1)(a_2 x - r_2)\ldots(a_n x - r_n). \qquad (15.2)$$

In Newton's day, the factorization might not have been quite as complete but rather had some quadratic factors remaining in order to avoid complex conjugate pairs. The idea is the same. By setting $P(x) = 0$ and solving, the roots of $P(x)$ are the values

$$\frac{r_1}{a_1}, \ldots, \frac{r_n}{a_n}.$$

Let r be the product $r_1 \ldots r_n$. Multiplying out the right-hand side of (15.2) and comparing with (15.1), $(-1)^n r_1 r_2 \ldots r_n = 1$. Thus,

$$r = r_1 r_2 \ldots r_n = (-1)^n. \qquad (15.3)$$

In addition,

$$P(x) = a_1 a_2 \ldots a_n x^n + \ldots + (-1)^{n-1}(a_1 r_2 \ldots r_n + a_2 r_1 r_3 \ldots r_n + \ldots$$
$$+ a_n r_1 \ldots r_{n-1})x + 1,$$

where each term in parentheses of the penultimate term is a product containing all factors $r_1, \ldots, r_n$ but one. Hence, the negative of the coefficient of the linear term

$$= -(-1)^{n-1} \left(\frac{a_1 r}{r_1} + \frac{a_2 r}{r_2} + \ldots + \frac{a_n r}{r_n} \right)$$

$$= (-1)^{2n} \left(\frac{a_1}{r_1} + \frac{a_2}{r_2} + \ldots + \frac{a_n}{r_n} \right) \text{ by using (15.3)}$$

$$= \frac{a_1}{r_1} + \frac{a_2}{r_2} + \ldots + \frac{a_n}{r_n},$$

the sum of the reciprocals of the roots. The result follows.

Let's turn our attention to the sine function. The function $f(x) = \sin x$ is defined and infinitely differentiable for all values of x. The roots (or zeros) of the sine function are all multiples of π. We will "apply" Newton's theorem to the function $f(x) = \sin x$ even though we don't exactly have a polynomial that represents the sine function. Rather than look for a polynomial, the best alternative is to find an appropriate power series representation. We now establish the Taylor series about $x = 0$ for the function $f(x) = \sin x$. (The special case of Taylor series centered about $a = 0$ are usually called *Maclaurin series* in honor of another early calculus investigator, the Scottish mathematician Colin Maclaurin (1698–1746).) We know that the Taylor series will actually equal $\sin x$ for all values of x. Begin by making a chart of the derivatives of $f(x) = \sin x$ up to order five:

$$f(x) = \sin x, f(0) = 0$$
$$f'(x) = \cos x, f'(0) = 1$$
$$f''(x) = -\sin x, f''(0) = 0$$
$$f'''(x) = -\cos x, f'''(0) = -1$$
$$f^{(4)}(x) = \sin x, f^{(4)}(0) = 0$$
$$f^{(5)}(x) = \cos x, f^{(5)}(0) = 1$$

Notice that the derivative of sine has a repeating pattern of length four. We have enough information for the entire Taylor series.

In general, the Taylor series about $x = 0$ for $\sin x$ is the following series valid for all x:

$$\sin x = 0 + \frac{1}{1!}x + \frac{0}{2!}x^2 + \frac{-1}{3!}x^3 + \frac{0}{4!}x^4 + \frac{1}{5!}x^5 + \ldots \text{ or}$$

$$\sin x = x - \frac{x^3}{3!} + \frac{x^5}{5!} - \frac{x^7}{7!} + \ldots = \sum_{n=1}^{\infty} \frac{(-1)^{n+1}x^{2n-1}}{(2n-1)!}. \qquad (15.4)$$

Next, let $x \neq 0$ be such that $\sin x = 0$. Then

$$0 = x - \frac{x^3}{3!} + \frac{x^5}{5!} - \frac{x^7}{7!} + \ldots.$$

Divide through by x to obtain

$$O = 1 - \frac{x^2}{3!} + \frac{x^4}{5!} - \frac{x^6}{7!} + \ldots.$$

If we let $z = x^2$, then

$$O = 1 - \frac{z}{3!} + \frac{z^2}{5!} - \frac{z^3}{7!} + \ldots.$$

Let

$$p(z) = 1 - \frac{z}{3!} + \frac{z^2}{5!} - \frac{z^3}{7!} + \ldots.$$

The negative of the coefficient of the linear term is $\frac{1}{3!} = \frac{1}{6}$.

Although $p(z)$ is a power series rather than a polynomial, if we apply Newton's theorem, the result is that

$$\frac{1}{6} = \sum_{n=1}^{\infty} \frac{1}{(n\pi)^2}. \qquad (15.5)$$

Because the roots of $p(z)$ are π^2, $(2\pi)^2$, $(3\pi)^2$ and so on, now multiply both sides of (15.5) by π^2:

$$\frac{\pi^2}{6} = \sum_{n=1}^{\infty} \frac{1}{n^2}. \qquad (15.6)$$

A numerical check based on previous calculations of the sum convinced Euler that his result was correct. However, he also realized that his "proof," though absolutely brilliant, was based on a theorem that had not been established for power series. Being the perfectionist that he was, soon he found an alternative proof considered completely rigorous. However, the analysis required goes beyond our treatise here. We leave that for another day.

Another stunning result of Euler's is the following which he discovered in 1737:

Proposition: *The number e is irrational.*

The proof requires the Maclaurin series (Taylor series about $x = 0$) of the function $f(x) = e^x$. Since the derivative of e^x is itself, the derivatives of all orders of e^x is e^x. Since $e^0 = 1$, the Maclaurin series for $f(x) = e^x$ is

$$1 + x + \frac{x^2}{2!} + \frac{x^3}{3!} + \ldots + \frac{x^n}{n!} + \ldots.$$

Similar to the sine function, it can be shown that the Maclaurin series, in fact, equals e^x for all x. So it's legitimate to write

$$e^x = \sum_{k=0}^{\infty} \frac{x^k}{k!} \text{ for all } x.$$

Letting $x = 1$,

$$e = \sum_{k=0}^{\infty} \frac{1}{k!} = 1 + 1 + \frac{1}{2} + \frac{1}{6} + \frac{1}{24} + \ldots,$$

which gives us a rapidly converging series for the number e.

By the way, it's easy to remember the initial part of its decimal expansion with a little knowledge of the seventh president of the United States, Andrew Jackson, who was first elected in 1828, served

two terms, and died in 1845. The number $e = 2.718281828459045\ldots$
Let

$$s_n = \sum_{k=0}^{\infty} \frac{1}{k!}$$

be the nth partial sum of the series. Now suppose, contrary to what we wish to show, that $e = \frac{a}{b}$ is a rational number with $b > 1$. For all $n > 1$,

$$0 < e - s_n = \sum_{k=n+1}^{\infty} \frac{x^k}{k!} <$$
$$\frac{1}{(n+1)!}\left[1 + \frac{1}{n+1} + \frac{1}{(n+1)^2} + \frac{1}{(n+1)^3} + \ldots\right].$$

But

$$1 + \frac{1}{n+1} + \frac{1}{(n+1)^2} + \frac{1}{(n+1)^3} + \ldots$$

is a convergent geometric series with $a = 1$ and $r = \frac{1}{n+1}$. It converges to $a/(1-r) = \frac{n+1}{n}$. It follows that

$$0 < e - s_n < \frac{1}{(n+1)!} \cdot \frac{n+1}{n} = \frac{1}{n!n} \text{ for all } n > 1.$$

Letting $n = b$: $0 < e - s_b < \frac{1}{b!b}$ or equivalently, $0 < b!\,(e - s_b) < \frac{1}{b}$. However, if $e = \frac{a}{b}$, then

$$b!\,(e - s_b) = b!\left(\frac{a}{b} + 1 + 1 + \frac{1}{2!} + \frac{1}{3!} + \ldots + \frac{1}{b!}\right)$$

is an integer.

But there are no integers strictly between 0 and 1. Hence, our assumption that e is rational is erroneous. Therefore, e is irrational!

Exercises

1. Find the Maclaurin series (Taylor series about $x = 0$) for $f(x) = \sin 2x$.
2. Find the Maclaurin series for $f(x) = e^{4x}$.
3. Find the Maclaurin series for $f(x) = x \cos x$.
4. Find the Taylor series for the function $f(x) = x^3$ about $x = 1$.
5. Find the Maclaurin series for $f(x) = \ln(1 + x)$.
6. Verify Newton's polynomial root theorem on the function

$$f(x) = -\frac{1}{8}x^3 + \frac{7}{8}x^2 - \frac{7}{4}x + 1.$$

Chapter 16

Special Sums of Fractions

Mathematics is the music of reason.

James Joseph Sylvester (1814–97)

In this chapter, we will learn about some special sums of fractions—both finite and infinite. We will learn about some of the first analytic expressions developed for the constant π. No longer was it necessary to inscribe and circumscribe a large circle with regular polygons of ever-increasing number of sides. But before we embark on our next topic, it is helpful to discuss a very useful result called the *arithmetic-geometric mean inequality*.

Let us begin with the simple observation that the square of any real number must be nonnegative. The square of 0 is 0 while the square of any other real number (positive or negative) is strictly positive. Hence, given any two real numbers x_1 and x_2, the quantity

$$(x_1 - x_2)^2 = x_1^2 - 2x_1x_2 + x_2^2 \geq 0.$$

It follows that $x_1^2 + x_2^2 \geq 2x_1x_2$ and

$$x_1^2 + 2x_1x_2 + x_2^2 = (x_1 + x_2)^2 \geq 4x_1x_2.$$

Dividing by 4, we have

$$\left(\frac{x_1 + x_2}{2}\right)^2 \geq x_1x_2.$$

If x_1 and x_2 are positive, then upon taking square roots of both sides

$$\frac{x_1 + x_2}{2} \geq \sqrt{x_1x_2}. \tag{16.1}$$

In other words, the arithmetic mean of two positive real numbers is greater than or equal to its geometric mean (with equality holding if and only if $x_1 = x_2$).

Formula (16.1) can be generalized to establish the *arithmetic-geometric mean inequality*: If $x_1, x_2, \ldots, x_n \geq 0$, then

$$\frac{x_1 + x_2 + \ldots + x_n}{n} \geq \sqrt[n]{x_1 x_2 \ldots x_n}. \qquad (16.2)$$

Again, equality holds if and only if $x_1 = x_2 = \ldots = x_n$. In formula (16.1), we established the special case where $n = 2$.

Now let's take a peek at some related sums of fractions to determine if they add up to being an integer. First, we consider the fraction x/y where x and y are integers. Clearly x/y is an integer if and only if $x|y$ (x divides y). Otherwise, x/y is a nonintegral rational number.

Taking our analysis one wee step further, consider the Diophantine equation

$$\frac{x}{y} + \frac{y}{x} = 1.$$

A Diophantine equation is an algebraic equation where either integers or rational solutions are sought but not other real solutions. The term honors the ancient Greek algebraist Diophantus of Alexandria (*c.*250 CE) who wrote a thirteen-chapter book *Arithmetica* chock full of such problems and solutions. In our equation above, we seek positive integers x and y that satisfy the equation. But here there is no solution. Simply note that if $x = y$, then the sum is 2. Further,

$$\frac{x}{y} \cdot \frac{y}{x} = 1.$$

If $x < y$, then $\frac{y}{x} > 1$ and if $x > y$, then $\frac{x}{y} > 1$. Since the terms are each positive, their sum must exceed 1.

Could the sum equal 2? That is, consider

$$\frac{x}{y} + \frac{y}{x} = 2.$$

Certainly, all integers for which $x = y$ satisfies the equation. Are there any other solutions?

The answer is no as can be seen by using inequality (16.1). Let $x_1 = \frac{x}{y}$ and $x_2 = \frac{y}{x}$. By (16.1),

$$\frac{x}{y} + \frac{y}{x} \geq 2$$

with equality only if $x = y$. There are no further solutions.

Moving along, for an arbitrary positive integer $n > 2$, let's consider the equation

$$\frac{x}{y} + \frac{y}{x} = n. \qquad (16.3)$$

Without loss of generality, suppose $x < y$. If $x|y$, then y/x is an integer and x/y is not an integer. Hence, their sum could not add up to any integer n. If x doesn't divide y, then there must be some prime number p with $p|x$ but p not dividing y. Being prime, p does not divide y^2 either. But then, since

$$\frac{x}{y} + \frac{y}{x} = \frac{x^2 + y^2}{xy},$$

$p|xy$ but p does not divide $x^2 + y^2$. Since p divides the denominator but not the numerator, the expression $\frac{x}{y} + \frac{y}{x}$ cannot be an integer. Therefore, equation (16.3) has no solutions for $n > 2$.

Mathematicians like to answer one question by asking several more. This way, like the real world, our mathematical universe continues to expand. Consider the Diophantine equation

$$\frac{x}{y} + \frac{y}{z} + \frac{z}{x} = n. \qquad (16.4)$$

This equation is carefully analyzed in the book *250 Problems in Elementary Number Theory* by the extremely prolific and celebrated Polish mathematician Wacław Sierpiński (1882–1969). In total, he published over 700 papers and 50 books. We recount a sliver of it here.

Letting

$$x_1 = \frac{x}{y}, x_2 = \frac{y}{z}, x_3 = \frac{z}{x}$$

and $n = 3$ in the arithmetic-geometric mean inequality (16.2) yields

$$\frac{x}{y} + \frac{y}{z} + \frac{z}{x} \geq 3.$$

Hence, there are no solutions for $n = 1$ or $n = 2$. For $n = 3$, there is the readily apparent solution $x = y = z$. In fact, this is the only set of solutions for $n = 3$.

Solutions to equation (16.4) for many values of n larger than 3 have been found, but even now, a full accounting has never been established. No one knows if there's even a solution when $n = 4$. The mathematician J. Browkin discovered the solutions

$$\frac{1}{2} + \frac{2}{4} + \frac{4}{1} = 5 \text{ and } \frac{2}{12} + \frac{12}{9} + \frac{9}{2} = 6.$$

Many other values of n have been solved including 9, 10, and all integers 13 through 21. Care to take a look?

Previously, we have dealt with many infinite *sums* of fractions. Now we turn our attention to a beautiful infinite *product* of fractions due to French mathematician François Viète for a ratio involving the constant π (Figure 16.1). Viète studied law, was a legal counselor to the parliament of Brittany, later a member of the parliament, and a

Figure 16.1 François Viète. Reproduced from Dr. Manuel, public domain, via Wikimedia Commons.

member of the royal privy council to both Henry III and Henry IV. During the War of Portuguese Succession (1580–3), Viète successfully deciphered a Spanish code containing 400 characters, hence giving France a substantial advantage. King Phillip II complained to the Pope that the French were employing magic against his country "contrary to the practice of the Christian faith."

Viète worked on mathematics during his leisure time. Along with Simon Stevin (1548–1620), he was among the first scholars to champion and popularize the use of decimal fractions as an alternative to sexagesimal ones. He was also one of the first mathematicians to work comfortably with all six trigonometric functions. We commence with the angle addition formula for the sine function:

$$\sin(A + B) = \sin A \cos B + \cos A \sin B.$$

Let $A = \theta$ and $B = \theta$ to derive the double-angle formula

$$\sin(2\theta) = 2 \sin \theta \cos \theta. \qquad (16.5)$$

It follows that

$$\sin \theta = 2 \sin\left(\frac{\theta}{2}\right) \cos\left(\frac{\theta}{2}\right).$$

Similarly, the angle addition formula for the cosine function is

$$\cos(A + B) = \cos A \cos B - \sin A \sin B.$$

Let $A = \theta$ and $B = \theta$ in the above formula:

$$\cos(2\theta) = \cos^2\theta - \sin^2\theta$$
$$= \cos^2\theta - \left(1 - \cos^2\theta\right),$$

resulting in the double-angle formula

$$\cos(2\theta) = 2\cos^2\theta - 1.$$

It follows that

$$\cos \theta = 2\cos^2\left(\frac{\theta}{2}\right) - 1$$

and hence,

$$\cos^2\left(\frac{\theta}{2}\right) = \frac{1 + \cos \theta}{2}.$$

Taking square roots and simplifying for $0 \le \theta \le \pi/2$, we get

$$\cos\left(\frac{\theta}{2}\right) = \sqrt{\frac{1 + \cos \theta}{2}}$$
$$\cos\left(\frac{\theta}{2}\right) = \frac{1}{2}\sqrt{2 + 2\cos \theta} \qquad (16.6)$$

It follows from (16.5) that

$$1 = \sin\tfrac{\pi}{2} = 2\sin\tfrac{\pi}{4}\cos\tfrac{\pi}{4}$$
$$= 2^2 \sin\tfrac{\pi}{8}\cos\tfrac{\pi}{8}\cos\tfrac{\pi}{4}$$
$$= 2^3\sin\tfrac{\pi}{16}\cos\tfrac{\pi}{16}\cos\tfrac{\pi}{8}\cos\tfrac{\pi}{4} \qquad (16.7)$$
$$\cdots$$
$$1 = 2^{n-1}\sin\tfrac{\pi}{2^n}\cos\tfrac{\pi}{2^n}\cos\tfrac{\pi}{2^{n-1}}\cdots\cos\tfrac{\pi}{8}\cos\tfrac{\pi}{4}.$$

Multiplying both sides of (16.7) by $\tfrac{2}{\pi}$

$$\frac{2}{\pi} = \frac{2^n}{\pi}\sin\frac{\pi}{2^n}\cos\frac{\pi}{2^n}\cos\frac{\pi}{2^{n-1}}\cdots\cos\frac{\pi}{8}\cos\frac{\pi}{4}.$$

Work from right to left and note that

$$\cos\frac{\pi}{4} = \frac{\sqrt{2}}{2}.$$

Applying formula (16.6) obtains

$$\frac{2}{\pi} = \frac{\sin\left(\frac{\pi}{2^n}\right)}{\pi/2^n}\cdot\frac{\sqrt{2}}{2}\cdot\frac{\sqrt{2+\sqrt{2}}}{2}\cdot\frac{\sqrt{2+\sqrt{2+\sqrt{2}}}}{2}\cdots.$$

Next, take the limit as n approaches infinity and let m $= \pi/2^n$:

$$\lim_{n\to\infty}\frac{\sin\left(\frac{\pi}{2^n}\right)}{\frac{\pi}{2^n}} = \lim_{m\to 0}+\frac{\sin m}{m} = 1.$$

Therefore,

$$\frac{2}{\pi} = \frac{\sqrt{2}}{2}\cdot\frac{\sqrt{2+\sqrt{2}}}{2}\cdot\frac{\sqrt{2+\sqrt{2+\sqrt{2}}}}{2}\cdots.$$

This is Viète's product formula. It ushered in a new age of analytical techniques to evaluate the constant π.

We conclude this chapter with a stunning infinite product representation for the number π. Unlike Viète's formula, this infinite product involves only *rational* numbers. It's the brainchild of the British mathematician John Wallis (1616–1702) (Figure 16.2).

Wallis was a true polymath, taking mathematics lessons under William Oughtred (1579–1660), the Cambridge University scholar who originate the symbol X for multiplication and : : for proportionality. Wallis studied medicine and devised a system for teaching the deaf. In 1649, he was appointed Savilian Professor of Geometry at Oxford University and held that position until his death. In addition to his academic duties of lecturing, scholarship, and administration, he also worked as a cryptanalyst for the government.

Figure 16.2 John Wallis, painted by Godfrey Kneller.
Reproduced from National Portrait Gallery, public domain, via Wikimedia Commons.

In 1656, Wallis published *Arithmetica Infintorum*, which clarified and extended the work of René Descartes and Bonaventura Cavalieri on analytic geometry and contained the seeds of calculus. Our modern symbol for infinity (∞) was first introduced here. He worked capably with negative and fractional exponents, new concepts at the time.

In 1662, along with the chemist Robert Boyle and the mathematician and architect Christopher Wren, Wallis helped form the Royal Society, to this day the pre-eminent scientific society of the United Kingdom. Later in his career, he discovered the formula for arclength of a curve which remains a key concept studied in college calculus courses. He applied his formula to both the cycloid (curve followed by a point on the outer edge of a circular wheel as it rolls horizontally along a straight line) and the curve $x^3 = ay^2$, known as a semi-cubical parabola. He also wrote an algebra textbook that included some of Newton's work on the binomial theorem.

The formula that Wallis discovered and included in the *Arithmetica Infinitorum* is

$$\frac{2}{\pi} = \frac{1}{2} \cdot \frac{3}{2} \cdot \frac{3}{4} \cdot \frac{5}{4} \cdot \frac{5}{6} \cdot \frac{7}{6} \cdots \qquad (16.8)$$

The proof requires some integral calculus, including the technique of integration by parts and the use of a reduction formula for integrating powers of the sine function. Integration by parts is the analog for integration that the product rule is for differentiation. Its main purpose is to replace a difficult integral with a more amenable one. The integration by parts formula states that

$$\int u \, dv = uv - \int v \, du.$$

Begin by defining the integral

$$I = \int \sin^n x \, dx. \qquad (16.9)$$

Let $u = \sin^{n-1}x$ and $dv = \sin x\, dx$. Then $du = (n-1)\sin^{n-2}x \cos x\, dx$ and $v = -\cos x$. Integrating by parts, we obtain

$$
\begin{aligned}
I &= -\cos x \sin^{n-1}x + (n-1)\int \sin^{n-2}x \, \cos^2x \, dx \\
&= -\cos x \sin^{n-1}x + (n-1)\int \sin^{n-2}x \left(1 - \sin^2x\right) dx \qquad (16.10) \\
&= -\sin^{n-1}x \cos x + (n-1)\int \sin^{n-2}x \, dx - (n-1)I.
\end{aligned}
$$

Collecting terms with factors of I and simplifying yields our reduction formula

$$
I = -\frac{1}{n}\sin^{n-1}x \, \cos x + \frac{n-1}{n}\int \sin^{n-2}x \, dx. \qquad (16.11)
$$

$$
\text{Let } I_n = \int_0^{\pi/2} \sin^n x \, dx \text{ for } n \geq 0. \qquad (16.12)
$$

Then

$$
I_0 = \int_0^{\pi/2} 1 \, dx = \frac{\pi}{2} \text{ and } I_1 = \int_0^{\pi/2} \sin x \, dx = -\cos x\big|_0^{\pi/2} = 1.
$$

Since $\sin 0 = 0$ and $\cos \pi/2 = 0$, from equations (16.11) and (16.12) we get

$$
I_k = \frac{k-1}{k}I_{k-2} \text{ for } k \geq 2. \qquad (16.13)
$$

Letting $k = 2n$ in (16.13), we obtain

$$
\begin{aligned}
I_{2n} &= \frac{2n-1}{2n}I_{2n-2} \\
&= \frac{2n-1}{2n} \cdot \frac{2n-3}{2n-2}I_{2n-4} \\
&= \frac{2n-1}{2n} \cdot \frac{2n-3}{2n-2} \cdot \frac{2n-5}{2n-4} \cdots \frac{1}{2}I_0 \\
I_{2n} &= \frac{1}{2} \cdot \frac{3}{4} \cdot \frac{5}{6} \cdot \frac{7}{8} \cdots \frac{2n-1}{2n} \cdot \frac{\pi}{2}. \qquad (16.14)
\end{aligned}
$$

Letting $k = 2n + 1$ in (16.13), we obtain

$$I_{2n+1} = \frac{2n}{2n+1} I_{2n-1}$$

$$= \frac{2n}{2n+1} \cdot \frac{2n-2}{2n-1} I_{2n-3}$$

$$= \frac{2n}{2n+1} \cdot \frac{2n-2}{2n-1} \cdot \frac{2n-4}{2n-3} \cdots \frac{2}{3} I_1$$

$$I_{2n+1} = \frac{2}{3} \cdot \frac{4}{5} \cdot \frac{6}{7} \cdot \frac{8}{9} \cdots \frac{2n}{2n+1}. \qquad (16.15)$$

Next, we need to determine

$$\lim_{n \to \infty} \frac{I_{2n+1}}{I_{2n}}.$$

The sine function satisfies the inequality $0 \leq \sin x \leq 1$ for $0 \leq x \leq \pi/2$. Thus,

$$0 \leq \sin^{2n+2}x \leq \sin^{2n+1}x \leq \sin^{2n}x \text{ for } 0 \leq x \leq \pi/2,$$

which implies that

$$0 < \int_0^{\pi/2} \sin^{2n+2}x \, dx \leq \int_0^{\pi/2} \sin^{2n+1}x \, dx \leq \int_0^{\pi/2} \sin^{2n}x \, dx$$

$$\text{i.e. } 0 < I_{2n+2} \leq I_{2n+1} \leq I_{2n}. \text{ Thus,}$$

$$0 < \frac{I_{2n+2}}{I_{2n}} \leq \frac{I_{2n+1}}{I_{2n}} \leq 1.$$

But (16.13) implies that

$$I_{2n+2} = \frac{2n+1}{2n+2} I_{2n}.$$

Hence,

$$\frac{2n+1}{2n+2} \leq \frac{I_{2n+1}}{I_{2n}} \leq 1.$$

Since

$$\lim_{n \to \infty} \frac{2n+1}{2n+2} = 1,$$

by the "squeeze theorem,"

$$\lim_{x \to \infty} \frac{I_{2n+1}}{I_{2n}} = 1.$$

Dividing equation (16.15) by equation (16.14) yields

$$\frac{I_{2n+1}}{I_{2n}} = \frac{2}{1} \cdot \frac{2}{3} \cdot \frac{4}{3} \cdot \frac{4}{5} \cdot \frac{6}{5} \cdot \frac{6}{7} \cdots \frac{2n}{2n-1} \cdot \frac{2n}{2n+1} \cdot \frac{2}{\pi}.$$

Thus,

$$\frac{\pi}{2} = \frac{2}{1} \cdot \frac{2}{3} \cdot \frac{4}{3} \cdot \frac{4}{5} \cdot \frac{6}{5} \cdot \frac{6}{7} \cdots \frac{2n}{2n-1} \cdot \frac{2n}{2n+1} \cdot \frac{I_{2n}}{I_{2n+1}}.$$

Taking the limit as $n \to \infty$, the Wallis product is established:

$$\frac{\pi}{2} = \frac{2}{1} \cdot \frac{2}{3} \cdot \frac{4}{3} \cdot \frac{4}{5} \cdot \frac{6}{5} \cdot \frac{6}{7} \cdots \frac{2n}{2n-1} \cdot \frac{2n}{2n+1} \cdots. \qquad (16.16)$$

Often the reciprocal is given instead:

$$\frac{2}{\pi} = \frac{1}{2} \cdot \frac{3}{2} \cdot \frac{3}{4} \cdot \frac{5}{4} \cdot \frac{5}{6} \cdot \frac{7}{6} \cdots \frac{2n-1}{2n} \cdot \frac{2n+1}{2n} \cdots. \qquad (16.17)$$

Since

$$\frac{1}{2} \cdot \frac{3}{2} = 1 - \frac{1}{2^2}, \quad \frac{1}{4} \cdot \frac{3}{4} \cdot \frac{5}{4}... $$
$$\frac{1}{2} \cdot \frac{3}{2} = 1 - \frac{1}{2^2}, \quad \frac{3}{4} \cdot \frac{5}{4} = 1 - \frac{1}{4^2},$$

and more generally

$$\frac{2n-1}{2n} \cdot \frac{2n+1}{2n} = 1 - \frac{1}{(2n)^2},$$

equation (16.17) can be rewritten as

$$\frac{2}{\pi} = \left(1 - \frac{1}{2^2}\right)\left(1 - \frac{1}{4^2}\right)\left(1 - \frac{1}{6^2}\right) \cdots \left(1 - \frac{1}{(2n)^2}\right) \cdots.$$

Exercises

1. What approximation to π do you obtain by letting $n = 10$ in the Wallis product (16.16)?

2. Knowing $\cos(\pi/4) = \sqrt{2}/2$ and $\cos(\pi/6) = \sqrt{3}/2$, use formula (16.6) to determine $\cos(\pi/8)$ and $\cos(\pi/12)$.

3. (a) Use the double-angle formula for sine (16.5) to show that
   $$\cos x = \frac{\sin 2x}{2 \sin x}.$$

   (b) Use part (a) to determine similar formulae for $\cos 2x$ and $\cos 4x$.

   (c) Show that
   $$\cos x \cos 2x \cos 4x = \frac{1}{8} \frac{\sin 8x}{\sin x}.$$

   (d) Let $x = \pi/9$ and utilize the fact that $\sin(\pi - x) = \sin x$ to show that

   $$\cos(\pi/9) \cos(2\pi/9) \cos(4\pi/9) = 1/8.$$

   (This was known as Morrie's Law by the American physicist Richard Feynman, who was shown this identity by his childhood friend Morrie Jacobs.)

4. (a) Establish the identity $\cos(\pi/17) \cos(2\pi/17) \cos(4\pi/17) \cos(8\pi/17) = 1/16$.

   (b) Show that

   $$\cos(\pi/9) \cos(2\pi/9) \cos(3\pi/9) \cos(4\pi/9) = \cos(\pi/17)$$
   $$\cos(2\pi/17) \cos(4\pi/17) \cos(8\pi/17).$$

Chapter 17

Some Spectacular Sums of Fractions

It is impossible to be a mathematician without being a poet in soul.

Sofya Kovalevskaya (1850–91)

If you've made it this far, you've worked hard and covered a lot of ground. Beginning with common fractions and basic decimals, you've engaged with Farey fractions, finite and infinite continued fractions, Egyptian fractions, infinite series, and various power series including Taylor series. We have had the pleasure of entering the minds of several great mathematicians from various countries and different epochs. Now I'd like to mention a couple more deep thinkers and courageous souls who have made our world more beautiful. The mathematical proofs of their discoveries go quite a bit beyond what we can recount here. In this, our final chapter, sit back, look out the window, check out the vista, and enjoy the ride!

Viscount William Brouncker (1620–84) was a close friend of John Wallis and for fifteen years was the first president of the Royal Society. The two men shared their mathematical investigations and discoveries. Brouncker made a fairly thorough analysis of what is now called Pell's equation, $x^2 - ny^2 = 1$, but unfortunately has not been afforded the historical recognition he deserved. Though usually considered an amateur mathematician, he also published some of his discoveries relating to the cycloid and other special curves. Brouncker was much impressed with Wallis's infinite product from

Chapter 16. He was inspired to rework it in the form

$$\frac{4}{\pi} = 1 + \cfrac{1^2}{2 + \cfrac{3^2}{2 + \cfrac{5^2}{2 + \cfrac{7^2}{2 + \cfrac{9^2}{2 + \dots}}}}}.$$

Though Brouncker did not provide a demonstration of his result, Wallis happily included it in his book *Arithmetica Infintorum*. Even knowing Wallis's formula, the result is not at all easy to establish. The first published proof of it that I am aware of is due to Euler (1776). We leave it for you to investigate at some point if you are so inclined.

We have mentioned Euler repeatedly in these pages. Among his numerous outstanding discoveries were a few continued fractions presented in his paper, "De fractionibus continuis dissertation" of 1737. The greatest of which is undoubtedly a simple contin-ued fraction for the constant e, the base of the natural logarithm function:

$$e = [2; 1, 2, 1, 1, 4, 1, 1, 6, 1, 1, 8, 1, 1, \dots]$$

$$= 2 + \cfrac{1}{1 + \cfrac{1}{2 + \cfrac{1}{1 + \cfrac{1}{1 + \cfrac{1}{4 + \cfrac{1}{1 + \cfrac{1}{1 + \cfrac{1}{6 + \dots}}}}}}}$$

Another of Euler's discoveries is this simple continued fraction:

$$\frac{e - 1}{e + 1} = [0; 2, 6, 10, 14, 18, \dots].$$

Euler also presented several interesting infinite sums for powers of π whose discovery he attributed to his friend, "the renowned" Chris-tian Goldbach (1690–1764). Of course, Goldbach is best remem-bered for his as-yet unproven conjecture (1742) that all even integers greater or equal to 4 are expressible as the sum of two primes.

Here are two of his identities:

$$\frac{\pi^4}{72} = \frac{1}{1^3} \cdot 1 + \frac{1}{2^3}\left(1 + \frac{1}{2}\right) + \frac{1}{3^3}\left(1 + \frac{1}{2} + \frac{1}{3}\right) + \ldots$$

as well as

$$\frac{\pi^4}{32} - \frac{\pi}{16} = \frac{1}{1^2 3^2} + \frac{1}{5^2 7^2} + \frac{1}{9^2 11^2} + \ldots.$$

We complete our tour of fractions by introducing a more modern and singularly brilliant individual, Srinivasa Ramanujan (1887–1920) (Figure 17.1).

Though a member of the Brahmin caste, Ramanujan grew up in severe poverty in southern India. He was largely self-taught; early on, his learning and insights far outstripped any human or material resources near at hand. His letters to mathematicians on "the Continent" were largely ignored. They contained long lists of

Figure 17.1 Srinivasa Ramanujan (1887–1920).

impossible-looking formulas without proof or explanation. But his nine-page letter in 1913 to the Cambridge analyst and number theorist G. H. Hardy hit its mark. Hardy was the rare mathematician who could perceive Ramanujan's brilliance. Hardy later wrote, "The [theorems] defeated me completely. I had never seen anything like them before." However, Hardy added that the formulas were so creative and fanciful that they must have emanated from a genius since no crank could possibly have the imagination to create them. Here are two relevant examples (the second includes the gamma function which is an important interpolation of the factorial function):

$$1 - 5\left(\frac{1}{2}\right)^3 + 9\left(\frac{1\cdot 3}{2\cdot 4}\right)^3 - 13\left(\frac{1\cdot 3\cdot 5}{2\cdot 4\cdot 6}\right)^3 + \ldots = \frac{2}{\pi}$$

$$1 + 9\left(\frac{1}{4}\right)^4 + 17\left(\frac{1\cdot 5}{4\cdot 8}\right)^4 + 25\left(\frac{1\cdot 5\cdot 9}{4\cdot 8\cdot 12}\right)^4 + \ldots = \frac{2^{3/2}}{\pi^{1/2}\{\Gamma\left(\frac{3}{4}\right)\}^2}.$$

With Hardy's help, Ramanujan traveled to England and spent several fruitful years working on his mathematics, occasionally with Hardy or with his colleague J. E. Littlewood. He made groundbreaking discoveries in analytic number theory, the theory of partitions, representations of numbers as sums of squares, infinite series, continued fractions, and the esoteric areas of elliptic and modular forms. In 1918, he was elected as a Fellow of the Royal Society and later that year a fellow of Trinity College, Cambridge. Unfortunately, Ramanujan's health deteriorated rapidly during this time. He returned to India in 1919 and died shortly thereafter. Commenting on Ramanujan, Einstein said, "I would not even attempt to understand how a prodigy's mind works. I will just remain jealous." Feel free to marvel at his "theorems" along with the rest of the mathematical community. The results included some continued fractions such as

$$\sqrt[4]{5}\sqrt{\frac{1+\sqrt 5}{2}} - \frac{1+\sqrt 5}{2} = \cfrac{e^{-2\pi/5}}{1 + \cfrac{e^{-2\pi}}{1 + \cfrac{e^{-4\pi}}{1 + \cfrac{e^{-6\pi}}{1 + \ldots}}}}$$

and

$$\frac{\sqrt{6\sqrt{3} - \left(1 + \sqrt{3}\right)}}{4} = \cfrac{e^{-2\pi/3}}{1 + \cfrac{e^{-2\pi} + e^{-4\pi}}{1 + \cfrac{e^{-4\pi} + e^{-8\pi}}{1 + \cfrac{e^{-6\pi} + e^{-12\pi}}{1 + \dots}}}}.$$

In 1914, Ramanujan published several rapidly converging series still of interest in calculating the digits of π:

$$\frac{1}{\pi} = \sum_{n=0}^{\infty} \frac{(2n)!^3 4^{2n+5}}{n!^6 2^{12n+4}}$$

and

$$\frac{1}{\pi} = \frac{\sqrt{8}}{9801} \sum_{n=0}^{\infty} \frac{(4n)!}{n!^4} \frac{1{,}103 + 26{,}390n}{396^{4n}}.$$

Enjoy learning more and making your own discoveries. As long as human beings create and wonder, mathematics will always flourish.

Exercises

1. Compare the following three approximations for e:
   (a) $e \approx 1 + 1 + \frac{1}{2} + \frac{1}{6} + \frac{1}{24} + \frac{1}{120} + \frac{1}{720} + \frac{1}{5040}$
   (b) $e \approx [2; 1, 2, 1, 1, 4, 1, 1, 6]$
   (c) $e \approx 1 + \frac{1}{2} + \frac{1}{3} + \frac{1}{4} + \frac{1}{5} + \frac{1}{6} + \frac{1}{7} + \frac{1}{8}.$

Exercise Solutions

Chapter 1

1. 0.0625, 0.025, 0.0016.
2. 0.1875, 3.275, 0.072.
3. $0.\overline{142857}$, $0.\overline{285714}$, $0.\overline{428571}$, $0.\overline{571428}$, $0.\overline{714285}$, $0.\overline{857142}$.
4. $0.\overline{076923}$, $0.\overline{153846}$, $0.\overline{230769}$, $0.\overline{307692}$, $0.\overline{384615}$, $0.\overline{461538}$, etc.
   Note the relationship between $\dfrac{n}{13}$ and $\dfrac{13-n}{13}$.
5. $\dfrac{3}{25}$, $\dfrac{49}{200}$, $\dfrac{60,001}{10,000}$.
6. $\dfrac{4}{33}$.
7. $\dfrac{34}{225}$.
8. 1882, $\dfrac{1}{3}$, $\dfrac{16}{45}$.
9. $\dfrac{11}{59}$.
10. $\dfrac{1}{3}$.
11. $a_2 = \dfrac{99}{70}$, $a_3 = \dfrac{19,601}{13,860} = 1.41421356\ldots$, an approximation for $\sqrt{2}$ accurate to eight decimal places.
12. 3; 47, 13, $20 = 3 + \dfrac{47}{60} + \dfrac{13}{3600} + \dfrac{20}{216,000}$.
13. There are $24 \times 60 \times 60 = 86,400$ seconds in a day. With the suggested decimal time system, there would be $10 \times 100 \times 100 = 100,000$ seconds in a day. Hence, each new decimal second would be 0.864 of a current second.
14. The probability that all three were born on different days of the week is

$$\frac{7}{7} \cdot \frac{6}{7} \cdot \frac{5}{7} = \frac{30}{49}.$$

So the probability at least two are born on the same day

$$= 1 - \frac{30}{49} = \frac{19}{49}.$$

The probability that none were born on a weekend is

$$\frac{5}{7} \cdot \frac{5}{7} \cdot \frac{5}{7} = \frac{125}{343}.$$

But

$$\frac{19}{49} = \frac{133}{343} > \frac{125}{343}.$$

So it's more likely that at least two were born on the same day of the week than that none were born on a weekend.

Chapter 2

1. (b), (d), (e) are geometric series.
2. (a) Put $2^n - 1$ dollars in wallet n for $1 <= n <= 7$.
   (b) Put $2^n - 1$ dollars in wallet n for $1 <= n <= 10$.
3. (a) $\sum_{n=1}^{\infty} \left(\frac{1}{10}\right)^n = \frac{1}{9}$, (b) $\sum_{n=1}^{\infty} \left(\frac{12}{100}\right)^n = \frac{4}{33}$, (c) $2 + \sum_{n=1}^{\infty} \left(\frac{34}{100}\right)^n = \frac{232}{99}$,
   (d) $\sum_{n=1}^{\infty} \left(\frac{123}{1000}\right)^n = \frac{41}{333}$.
4. $0.137174211248285322\ldots$
5. $20_3 = \frac{8}{9}$ is in C while $12_3 = \frac{5}{9}$ is not in C, for example.
6. (c) $\sum_{n=1}^{\infty} \left(\frac{1}{9}\right)\left(\frac{8}{9}\right)^{n-1} = 1$ ($a = 1/9, r = 8/9$).
7. 84 years.

Chapter 3

1. $1, 1, 2, 3, 5, 8, 13, 21, 34, 55, 89, 144, 233, 377, 610$. The numbers 1 and 144 are squares. It's been shown there are no more Fibonacci squares.
3. If $x > 0$ and $\frac{1}{x} = \frac{x}{1-x}$, then $x^2 + x - 1 = 0$. Quadratic formula implies that $x = \frac{-1+\sqrt{5}}{2}$. In this case,

$$\frac{1}{x} = \frac{1 + \sqrt{5}}{2} = \varphi.$$

4. Use induction and fact $f_{n+1} + f_{n+2} = f_{n+3}$.
5. Use induction.
6. Use induction.
7. Use induction.
8. Use induction. The identity holds for the case $n = 1$. Assume it holds for case n. Then

$$f_{n+1}f_{n-1} - f_n^2 + f_n f_{n+1} - f_n f_{n+1} = (-1)^n.$$

Hence,

$$f_{n+1}(f_n + f_{n-1}) - f_n(f_n + f_{n+1}) = (-1)^n.$$

So

$$f_{n+1}^2 - f_n f_{n+2} = (-1)^n.$$

Multiplying through by -1 establishes the proposition for

$$n+1: f_n f_{n+2} - f_{n+1}^2 = (-1)^{n+1}.$$

9. $0.0001010203050813213455\ldots$.
10. $60/37$ hours.
11. The total amount is 282 units of money. They took 198, 78, and 6 units initially. (The problem is actually indeterminate, but this is the smallest positive solution.)
12. (a) $73 = 8^2 + 3^2$.
    (b) $101 = 10^2 + 1^2$.
    (c) $7373 = 83^2 + 22^2 = 77^2 + 38^2$.
13. (a) Use induction. $f_0 + f_2 = 0 + 1 = L_1, f_1 + f_3 = 1 + 2 = L_2$. Assume true for all values up to some n. Then $f_{n-2} + f_n = L_{n-1}$ and $f_{n-1} + f_{n+1} = L_n$. Hence,

$$f_n + f_{n+2} = (f_{n-2} + f_{n-1}) + (f_n + f_{n+1}) = (f_{n-2} + f_n) + (f_{n-1} + f_{n+1})$$

$$= L_{n-1} + L_n = L_{n+1}.$$

   (b) $L_0 + L_2 = 2 + 3 = 5 = 5f_1. L_1 + L_3 = 1 + 4 = 5 = 5f_2$.
       Assume true for all values up to some n. Then $L_{n-2} + L_n = 5f_{n-1}$ and $L_{n-1} + L_{n+1} = 5f_n$. Hence,

$$L_n + L_{n+2} = (L_{n-2} + L_{n-1}) + (L_{n-1} + L_n)$$

$$= (L_{n-2} + L_n) + (L_{n-1} + L_{n+1})$$

$$= 5f_{n-1} + 5f_n = 5(f_{n-1} + f_n) = 5f_{n+1}.$$

Chapter 4

1. (a), (b), (c).
2. $x \equiv 2 \pmod 8$.
3. (a) $x \equiv 2 \pmod 7$, (b) $x \equiv 2, 5, 8 \pmod 9$.
5. $x \equiv 1, 3, 5, 7 \pmod 8$.
6. (a) If x is even with $x \equiv 0$ or $2 \pmod 4$, then $x^2 \equiv 0 \pmod 4$. If x is odd with $x \equiv 1$ or $3 \pmod 4$, then $x^2 \equiv 1 \pmod 4$.
   (b) All of the numbers are $\equiv 3 \pmod 5$. But squares can only be $\equiv 0, 1,$ or $4 \pmod 5$.
7. $10^{100} \equiv 1^{100} \equiv 1 \pmod 9.10^{100} \equiv (-1)^{100} \equiv 1 \pmod{11}$.
8. Use the fact that $24 = 2^3 \cdot 3$ and reason modulo 2, 3, and 4.

9. The number of zeros at the tail end of 1000! is determined by the power of 5 that divides 1000!. The answer is $\left[\frac{1000}{5}\right] + \left[\frac{1000}{25}\right] + \left[\frac{1000}{125}\right] + \left[\frac{1000}{625}\right] = 249$. Here $[x]$ is the greatest integer less than or equal to x.

10. Note that $10^{2n} \equiv 1 \pmod{11}$ and $10^{2n-1} \equiv -1 \pmod{11}$. For example, $2497 \equiv 1(7) + (-1)4 + 1(9) + (-1)2 \equiv 0 \pmod{11}$.

11. 69.

12. 727.

13. Consider $n^5 \pmod{10}$ for $n = 1, 2, \ldots, 10$.

14. (a) Symmetric, (b) transitive, (c) reflexive and transitive, (d) all— equivalence relation.

15. The Germain primes below 100 are 2, 3, 5, 11, 23, 29, 41, 53, 83, and 89. It is unknown if there are infinitely many of them in all.

Chapter 5

1. (a) 3, (b) 2, (c) 1.

2. (a) $1 \cdot 111 - 3 \cdot 36 = 3$, (b) $13 \cdot 16 - 1 \cdot 206 = 2$, (c) $34 \cdot 144 - 55 \cdot 89 = 1$.

3. (a) 1332, (b) 1648, (c) 12,816.

4. (a) $\gcd(m, n) = 1 \Rightarrow$ there are x, y such that $mx + ny = 1$. Hence, $dmx + dny = d$.
   (b) $x = 80, y = -60$.

5. Bezout $\Rightarrow$ there exist x_0, y_0 with $ax_0 + by_0 = 1$. But then $x = x_0 + bn, y = y_0 - an$ are solutions for all n.

6. $(p, h) = (4, 11), (12, 8), (20, 5),$ and $(28, 2)$.

7. $17^* \equiv -6 \equiv 97 \pmod{103}$. $x \equiv -24 \equiv 79 \pmod{103}$.

8. $71^* \equiv 43 \pmod{109}$. $x \equiv 20 \pmod{109}$.

9. S must contain two consecutive integers.

10. (a) 87, (b) 100.

Chapter 6

1. (a) 20, (b) 72, (c) 48, (d) 2026, (e) 400,000,000.

3. Must have a single representative from each residue class.

5. Must have a single representative from each residue class relatively prime to 7.

6. 3 divides 111 and all strings of 1s containing a multiple of 3 number of 1's. If $p > 5$, then $\gcd(10, p) = 1$ and $10^{p-1} \equiv 1 \pmod{p}$. So p divides a string of $p - 1$ 9's or any multiple of $p - 1$ 9's. Now divide by 9.

7. (a) $0.\overline{076923}$.

8. (a) $0.\overline{052631578947368421}\ldots$

9. Reason modulo 10.

10. $2^4 \equiv 1 \pmod{10}$ and so $2^{1,000,000} = \left(2^4\right)^{250,000} \equiv 1^{250,000} \equiv 1 \pmod{10}$.

11. Consider the n fractions $1/n, 2/n, \ldots, n/n$ reduced to lowest terms.

12. 7, 17, 19, 23, 29, 47, 59, 61, 97.

13. Note the relationship between $n/13$ and $(13 - n)/13$. The twelve fractions break into two groups of six fractions, each containing the same repeated pattern of six digits. The first group contains the fractions 1/13, 10/13, $10^2/13$, $10^3/13$ where we simplify $10^n.(\text{mod} \cdot 13)$ as we proceed. The second group contains 2/13, 20/13, $2 \cdot /10^2/13$, etc.

14. The thirty-six fractions break into twelve separate groups of three fractions, each with the same repeated pattern of three digits. Each group contains $a/37$, $10a/37$, and $(37 - 11a)/37$ with the numerator simplified modulo 37.

Chapter 7

1. (a) $\begin{bmatrix} 4 & 3 \\ 5 & 2 \end{bmatrix}$, (c) $\begin{bmatrix} 5 & -4 & 5 \\ -1 & 0 & -3 \end{bmatrix}$, (f) $\begin{bmatrix} 0 & -2 & -5 \\ -11 & 4 & -23 \end{bmatrix}$.

2. $\dfrac{1}{2} \begin{bmatrix} 0 & -2 \\ 1 & 1 \end{bmatrix}$.

3. $\begin{bmatrix} -6 & -2 \\ 9 & 3 \\ 2 & 2 \end{bmatrix}$.

4. $\det A = -1$, $\det B = -7$, $\det C = 0$, $\det D = 1$.
$A^{-1} \begin{bmatrix} -5 & 2 \\ 3 & -1 \end{bmatrix}$, $B^{-1} = -\dfrac{1}{7} \begin{bmatrix} 4 & -3 \\ -5 & 2 \end{bmatrix}$, $D^{-1} = \begin{bmatrix} 1 & 0 \\ 0 & 1 \end{bmatrix}$.

Chapter 8

1. (a) F_7 : 0/1, 1/7, 1/6, 1/5, 1/4, 2/7, 1/3, 2/5, 3/7, 1/2, 4/7, 3/5, 2/3, 5/7, 3/4, 4/5, 5/6, 6/7, 1/1.

   (b) F_8 : 0/1, 1/8, 1/7, 1/6, 1/5, 1/4, 2/7, 1/3, 3/8, 2/5, 3/7, 1/2, 4/7, 3/5, 5/8, 2/3, 5/7, 3/4, 4/5, 5/6, 6/7, 7/8, 1/1.

3. 7/5.

4. 9/4.

5. 20/9.

6. 37/5.

7. (a) $\dfrac{a}{b} = \dfrac{7}{14}$, $\dfrac{c}{d} = \dfrac{2}{6}$, (b) $\dfrac{c}{d} = \dfrac{2}{6}$, $\dfrac{c}{d} = \dfrac{1}{2}$.

8. (a) $\dfrac{a}{b} = \dfrac{1}{4}$, $\dfrac{c}{d} = \dfrac{2}{6}$, (b) $\dfrac{a}{b} = \dfrac{3}{12}$, $\dfrac{c}{d} = \dfrac{4}{12}$.

Chapter 9

1. Maximum number of pieces is

$$1 + (1 + 2 + \ldots + n) = \frac{n^2 + n + 2}{2}.$$

$2n$ of the pieces have some crust.

2. (a) Put 2^{n-1} dollars in wallet n for $1 < n < 7$.

   (b) Put Put 2^{n-1} dollars in wallet n for $1 < n < 10$.

3. Use the Pythagorean theorem.

4. After the first break, we have an equal chance of choosing the smaller or larger one to snap next. If we choose the smaller piece, then no division of it will result in a possible triangle since the sum of the two smaller pieces won't add up to the length of the third. In Figure 9.1, let PE represent the length of the smaller piece (of length $h/2$ or less). If we choose the larger piece to break further, then the possible positions for P range uniformly over the triangles $\triangle GBH$, $\triangle GIH$, and $\triangle HIC$ since P cannot be in $\triangle AGI$. But those that form a triangle still reside solely in $\triangle GIH$. So there is a 1/3 chance of making a triangle from the sticks given that we choose the larger piece to break further. The overall probability is then $\frac{1}{2} \cdot 0 + \frac{1}{2} \cdot \frac{1}{3} = \frac{1}{6}$.

5. If Alice has proportion a of the cake, Bob has proportion b, then Clair has proportion $1 - a - b$. By choosing the largest of the quartered pieces from each of them, Dan thinks he has at least $a/4 + b/4 + (1 - a - b)/4 = 1/4$ of the cake.

6. Cut the brick so that the areas of the pieces are 1/31, 2/31, 4/31, 8/31, and 16/31 of the area of the entire brick. Now appropriate pieces can be used to pay any integral amount from 1 day to 31 days.

7. In binary, the numbers 1 through 7 are written as 001, 010, 011, 100, 101, 110, and 111, respectively. Looking from right to left, the bits represent the numbers 1, 2, 4. A one represents including that bit while a zero means to exclude that bit. Similarly, just four bits are needed to represent the integers 1 through 15 and five bits to represent the integers 1 through 31.

Chapter 10

1. $\frac{2}{5} = \frac{1}{3} + \frac{1}{15}, \frac{3}{10} = \frac{1}{4} + \frac{1}{20}, \frac{2}{11} = \frac{1}{6} + \frac{1}{66}, \frac{6}{13} = \frac{1}{3} + \frac{1}{8} + \frac{1}{312}.$

2. $\frac{7}{12} = \frac{1}{3} + \frac{1}{4} = \frac{1}{2} + \frac{1}{11} + \frac{1}{132}.$

3. $\frac{4}{2} = \frac{1}{1} + \frac{1}{2}, \frac{1}{2}, \frac{4}{3} = \frac{1}{1} + \frac{1}{4} + \frac{1}{12}, \frac{4}{4} = \frac{1}{3} + \frac{1}{3} + \frac{1}{3}, \frac{4}{5} = \frac{1}{2} + \frac{1}{5} + \frac{1}{10}, \frac{4}{6} = \frac{1}{2} + \frac{1}{12} + \frac{1}{12}, \frac{4}{7} = \frac{1}{2} + \frac{1}{28} + \frac{1}{28}, \frac{4}{8} = \frac{1}{6} + \frac{1}{6} + \frac{1}{6}, \frac{4}{9} = \frac{1}{3} + \frac{1}{18} + \frac{1}{18}, \frac{4}{10} = \frac{1}{3} + \frac{1}{30} + \frac{1}{30}, \frac{4}{11} = \frac{1}{3} + \frac{1}{66} + \frac{1}{66}, \frac{4}{12} = \frac{1}{9} + \frac{1}{9} + \frac{1}{9}, \frac{4}{13} = \frac{1}{4} + \frac{1}{26} + \frac{1}{52}.$

5. $n_1 = 1, n_2 = 2, \ldots, n_{10} = 10$. Greedy algorithm works here.

6. (a) $\frac{1}{15} + \frac{1}{10} = \frac{1}{6}$.

   (b) $\frac{1}{28} + \frac{1}{21} = \frac{1}{12}$.

7. (a) $\dfrac{1}{156^2} + \dfrac{1}{65^2} = \dfrac{1}{60^2}$.

   (b) $\dfrac{1}{255^2} + \dfrac{1}{136^2} = \dfrac{1}{120^2}$.

8. For example, $a = 2$, $b = 5$ leads to $20^2 + 21^2 = 29^2$.

9. $24^2 + 7^2 = 25^2$.

10. Let P be a nonempty set of primes and $p \in P$. Rewrite the sum of reciprocals of all the primes in P by getting a common denominator which must be the product of all the primes in P. If the sum is d/n, then $d = n$ if the sum were 1. The prime p divides n. But p divides all terms in numerator n but one (the single term consisting of the product of all primes in P excepting p itself). So p does not divide d. Hence, $d \neq n$, contradicting the assumption that the sum of reciprocals was 1. This problem demonstrates that an infinite set of integers can have the property that none of its subsets have: a sum of reciprocals adding to 1.

11. The product 72×59 can be tallied as such:

72	59
36	118
18	236
9	472
4	944
2	1888
1	3776

    Hence, $72 \times 59 = 3776 + 472 = 4248$. To understand why this works, consider writing 72 in base 2 and then use the fact that multiplication distributes over addition.

Chapter 11

1. (a) Assume $\sqrt{3} = \frac{a}{b}$ and then mimic Euclid's proof for $\sqrt{2}$ or use the fundamental theorem of arithmetic.

   (b) $\sqrt{4} = \frac{a}{b} \Rightarrow a^2 = 4b^2 \Rightarrow a$
       a is even. But this doesn't imply that b is even.

2. Assume $\log_{10} 2 = \frac{a}{b}$. Then $10^{a/b} = 2 \Rightarrow 10^a = 2^b$. Now apply the fundamental theorem of arithmetic.

3. $1 + 3 + \ldots + (2n - 1) = n^2$. Use induction.

4. $21 = t_6$, $2211 = t_{66}$, $222{,}111 = t_{666}$, etc.

5. $41{,}616 = 204^2 = t_{288}$.

6. $\dfrac{n(3n-1)}{2}$.

7. (a) 24th position, 130th position.
   (b) 10/36.

8. Note that $\sqrt{2}^{\sqrt{2}}$ must be either rational or irrational!

9. The number is irrational. If it were rational, then there would eventually be some repeating pattern of digits of some set length (say n). But eventually

the digits include $n + 1$ zeros once we reach the integer 10^{n+1}. This number is known as both Champernowne's constant (named after D. G. Champernowne, who as an undergraduate in 1933 wrote about its irrationality), or as Mahler's number since the mathematician Kurt Mahler proved in 1937 that it is actually a transcendental number). Transcendental numbers are those that are not the root of any nonzero polynomial having rational coefficients.

Chapter 12

1. (a) $[5; 1, 7, 2]$, (b) $[0; 20, 5]$, (c) $[0; 1, 1, 1, 1, 1, 1, 1, 1, 1, 2]$.

2. $\dfrac{37}{29}$.

3. $[365; 4, 7, 1, 3, 4, 1, 1, 1, 2]$.

4. $\dfrac{365}{1}, \dfrac{1461}{4}, \dfrac{10{,}592}{29}, \dfrac{12{,}053}{33}$.

5. $[5; 5, 10, 10] = \dfrac{2676}{515} \approx 3\sqrt{3}$ with an error of about 0.000036.

6. $\sqrt{7} = [2, \overline{1, 1, 1, 4}]$.

7. (a) $[2; 4, 4, \overline{4}]$, (b) $[3; 6, 6, \overline{6}]$, (c) $[4; 8, 8, \overline{8}]$, (d) $[k; 2k, 2k, \overline{2k}]$.

8. If $x = \sqrt{1 + \sqrt{1 + \sqrt{1 + \ldots}}}$, then $x = \sqrt{1 + x}$. It follows that $x^2 = 1 + x$, and so $x = \varphi$.

   It follows that $x^2 = 1 + x$, and so $x = \varphi$.

9. Since x is irrational, so is qx for any positive integer q. Let p be the nearest integer to qx. So

$$-\frac{1}{2} < qx - p < \frac{1}{2}.$$

with strict inequalities since $qx - p$ is irrational. Dividing through by q gives

$$\frac{-1}{2q} < x - \frac{p}{q} < \frac{1}{2q}.$$

Add $\frac{p}{q}$ throughout to obtain

$$\frac{p}{q} - \frac{1}{2q} < x < \frac{p}{q} + \frac{1}{2q}.$$

In fact, it can be shown that given any irrational number x and any positive integer k, there is a rational number p/q with $q \le k$ such that

$$\frac{p}{q} - \frac{1}{kq} < x < \frac{p}{q} + \frac{1}{kq}.$$

10. (a) $p = 14$, (b) $p = 14$, (c) $p = 22$, (d) $p = 38$.

Chapter 13

1. $3 \cdot 2^{10}$.
2. $3 \cdot 2^{17}$.
3. $3 . 1416$.

Chapter 14

1. (a) $R = 1$, (b) $R = 2$, (c) $R = \frac{1}{2}$, (d) $R = 1$.
2. Rewrite

$$S_n = 1 + \frac{1}{2} + \ldots + \frac{1}{n} = \frac{a_1 + a_2 + \ldots + a_n}{n!}$$

1. where $a_i = \frac{n!}{i}$ for $1 \le i \le n$. By Bertrand's Postulate there is a prime p with $n/2 < p < n$. Then p divides the denominator and every term in the numerator save for ap. So p does not divide the numerator. Reason why sn cannot be an integer.
3. (a) 2, (b) 4/3.
4. (a) Ratio Test.
   (b) $\frac{a}{1-r} = \frac{x/2}{1-x/2} = \frac{x}{2-x}$.
   (c) Use quotient rule for differentiation.
   (d) $\sum_{n=1}^{\infty} \frac{n}{2^n} = 2$.
   (e) Use quotient rule for differentiation and then let $x = 1$.
   (f) $\sum_{n=1}^{\infty} \frac{n^2}{2^n} = 6$.
5. (a) $\sum_{n=1}^{\infty} \frac{n^2 x^n}{2^n} = \frac{2x^2+4x}{(2-x)^3}$.
   (b) $\sum_{n=1}^{\infty} \frac{n^3 x^{n-1}}{2^n} = \frac{2x^2+16x+8}{(2-x)^4}$.
   (c) $\sum_{n=1}^{\infty} \frac{n^8}{2^n} = 26$.
6. (a) Let $H_k = 1 + \frac{1}{2} + \ldots + \frac{1}{k}$ be the kth harmonic number. Then

$$1 - \frac{1}{2} + \frac{1}{3} - \ldots + \frac{1}{2n-1} - \frac{1}{2n} = H_{2n} - 2\left(\frac{1}{2} + \frac{1}{4} + \frac{1}{6} + \ldots + \frac{1}{2n}\right)$$

$$= H_{2n} - H_n = \frac{1}{n+1} + \frac{1}{n+2} + \ldots + \frac{1}{2n}.$$

(b) This is a variation on a problem from the 1970 International Mathematics Olympiad. The key idea is to pair up terms appropriately. Let

$$\frac{m}{n} = 1 - \frac{1}{2} + \frac{1}{3} - \frac{1}{4} + \ldots + \frac{1}{59}$$

$$= \left(1 - \frac{1}{2} + \frac{1}{3} - \frac{1}{4} + \ldots - \frac{1}{58}\right) + \frac{1}{59}$$

$$= \left(\frac{1}{30} + \ldots + \frac{1}{58} \right) + \frac{1}{59}$$

$$= \left(\frac{1}{30} + \frac{1}{59} \right) + \left(\frac{1}{31} + \frac{1}{58} \right) + \ldots + \left(\frac{1}{44} + \frac{1}{45} \right)$$

$$= \frac{30 + 59}{30 \cdot 59} + \frac{31 + 58}{31 \cdot 58} + \ldots + \frac{44 + 45}{44 \cdot 45} = \frac{89}{30 \cdot 59} + \ldots + \frac{89}{44 \cdot 45}$$

$$= \frac{89d}{30 \cdot 31 \cdot \ldots \cdot 59},$$

where d is some large integer. But 89 is prime and hence is relatively prime to the denominator. Hence, 89 divides m.

Chapter 15

1. $\sin 2x = \sum_{n=0}^{\infty} \frac{(-1)^n 2^{2n+1} x^{2n+1}}{(2n)!}$
2. $e^{4x} = \sum_{n=0}^{\infty} \frac{4^n x^n}{n!}$
3. $x \cos x = \sum_{n=0}^{\infty} \frac{(-1)^n x^{2n+1}}{(2n)!}$
4. $1 + 3(x - 1) + 3(x - 1)^2 + (x - 1)^3$
5. $\ln(1 + x) = \sum_{n=1}^{\infty} (-1)^{n+1} \frac{x^n}{n}$
6. $-\frac{1}{8}x^3 + \frac{7}{8}x^2 - \frac{7}{4}x + 1 = -\frac{1}{8}(x - 1)(x - 2)(x - 4)$.
   Sum of reciprocals of roots are $\frac{1}{1} + \frac{1}{2} + \frac{1}{4} = \frac{7}{4}$.

Chapter 16

1. $\frac{\pi}{2} \approx \frac{2}{1} \cdot \frac{2}{3} \ldots \frac{20}{19} \cdot \frac{20}{21} \approx 3.0677$
2. $\cos \frac{\pi}{8} = \frac{1}{2}\sqrt{2 + \sqrt{2}}, \cos \frac{\pi}{12} = \frac{1}{2}\sqrt{2 + \sqrt{3}}$
3. (b) $\cos 2x = \frac{\sin 4x}{2 \sin 2x}, \cos 4x = \frac{\sin 8x}{2 \sin 4x}$
   (d) $\cos \frac{\pi}{9} \cos \frac{2\pi}{9} \cos \frac{4\pi}{9} = \frac{1}{8} \frac{\sin \frac{8\pi}{9}}{\sin \frac{\pi}{9}} = \frac{1}{8}$.
   This is often written in terms of degrees rather than radians.
4. (a) $\cos x \cos 2x \cos 4x \cos 8x = \frac{1}{16} \frac{\sin 16x}{\sin x}$.
   Let $x = \frac{\pi}{17}$ and note $\sin \frac{\pi}{17} = \sin \frac{16\pi}{17}$.

Then

$$\cos \frac{\pi}{17} \cos \frac{2\pi}{17} \cos \frac{4\pi}{17} \cos \frac{8\pi}{17} = \frac{1}{16}.$$

(a) Use $\cos\left(\dfrac{\pi}{3}\right) = \dfrac{1}{2}$.

Chapter 17

1. (a) $\frac{685}{252} = 2.71825396825\ldots$, (b) $\frac{1264}{465} = 2.71827956989\ldots$, (c) $\frac{761}{280} = 2.717857142857\ldots$.

Bibliography

Aaboe, A. *Episodes from the Early History of Mathematics*. Mathematical Association of America, 1964.

Ayoub, R. "Euler and the Zeta Function." *American Mathematical Monthly*, vol. 81, 1974: 1067–86.

Baillie, R. "Sums of Reciprocals of Integers Missing a Given Digit." *American Mathematical Monthly*, vol. 86, 1979: 372–74.

Behforooz, G. H. "Thinning Out the Harmonic Series." *Math. Magazine*, vol. 68, 1995: 289–93.

Boyer, C. and Merzbach, U. C. *A History of Mathematics*. 2nd ed. Wiley, 1989.

Dickson, L. E. *History of the Theory of Numbers*. 3 vols. Chelsea, 1966.

Diophantus. *Arithemetica*. Trans. and commentary by J. Christianidis and J. Oaks. Rutledge, 2022.

Euclid. *The Thirteen Books of Euclid's Elements*. Trans. Heiberg with commentary and introduction by T. L. Heath. Dover, 1956.

Gardner, M. *The Scientific American Book of Mathematical Puzzles and Diversions*. Simon and Schuster, 1959.

Hardy, G. H. *Ramanujan*. 3rd ed. Chelsea, 1978.

Hardy, G. H. and Wright, E. M. *An Introduction to the Theory of Numbers*. 5th ed. Clarendon Press, 1979.

Katz, V. *A History of Mathematics—An Introduction*. 2nd ed. Addison Wesley, 1998.

Konhauser, J. D. E., Velleman, D., and Wagon S. *Which Way Did the Bicycle Go?* Mathematical Association of America, 1996.

LeVeque, W. J. *Fundamentals of Number Theory*. Addison-Wesley, 1977.

Maynard, J. "Primes with Restricted Digits." Inventiones Mathematicae, vol. 217, 2019: 127–18.

Niven, I. *Numbers: Rational and Irrational*. Mathematical Association of America, 1961.

Olds, C. D. *Continued Fractions*. Mathematical Association of America, 1956.

Pegg, E., Jr. "The Loculus of Archimedes, Solved," Nov. 17, 2003, http://www.maa.org/editorial/mathgames/mathgames_11_17_03.html

Schumer, P. D. *Introduction to Number Theory*. PWS, 1995.

Schumer, P. D. *Mathematical Journeys*. Wiley, 2004.

Schwartzman, S. *The Words of Mathematics: An Etymological Dictionary of Mathematical Terms Used in English*. Mathematical Association of America, 1994.

Sierpiński, W. *Elementary Theory of Numbers*. Pantswowe Wydawnictwo Naukowe, 1964.

Sierpiński, W. *250 Problems in Elementary Number Theory*. Elsevier, 1970.

Smith, D. E. *History of Mathematics.* 2 vols. Dover, 1951.

Stillwell, J. *Mathematics and its History.* 3rd ed. Springer, 2010.

Young, R. M. *Excursions in Calculus: An Interplay of the Continuous and the Discrete.* Mathematical Association of America, 1992.

Index